Pelican Books

Writing Technic

Bruce Cooper was born in Shrewsbury in
1925 and educated at Ratcliffe College,
Edinburgh University, and Peterhouse,
Cambridge. He served during the latter half
of the war with the Royal Artillery in
India. His first interest was in adult education,
and shortly after leaving Cambridge he
took up an appointment with the W.E.A.
as Tutor-Organizer for Suffolk, during
which time he also lectured to H.M. Forces
for Cambridge University Extra-Mural
Board. In 1956 he entered technical education,
joining the staff of the Hatfield College
of Technology, where he later became a
lecturer in Communication. During his
time there he was seconded to industry,
working with Esso Petroleum and helping
the de Havilland Aircraft Company develop
a supervisory training scheme. He became
interested in report writing, a subject in
which he has conducted many classes. In
1961 he accepted the post of Head of the
Department of Liberal Studies at
Stockton/Billingham Technical College.
His present appointment, which he took up
in 1966, is Management Training Officer for
the Agricultural Division of I.C.I.
He writes frequently for the educational press
and literary periodicals and contributes
regularly for the B.B.C. He is married to a
Swedish teacher and has five children.

Writing
Technical Reports

Bruce M. Cooper

Penguin Books

Penguin Books Ltd, Harmondsworth, Middlesex, England
Penguin Books, 625 Madison Avenue, New York, New York 10022, U.S.A.
Penguin Books Australia Ltd, Ringwood, Victoria, Australia
Penguin Books Canada Ltd, 2801 John Street, Markham, Ontario, Canada L3R 1B4
Penguin Books (N.Z.) Ltd, 182–190 Wairau Road, Auckland 10, New Zealand

First published 1964
Reprinted 1965, 1967, 1969, 1971, 1975, 1976, 1978, 1979, 1981

Copyright © Bruce M. Cooper, 1964
All rights reserved

Made and printed in Great Britain
by Hazell Watson & Viney Ltd, Aylesbury, Bucks
Set in Intertype Times

Except in the United States of America, this book is sold subject
to the condition that it shall not, by way of trade or otherwise, be lent,
re-sold, hired out, or otherwise circulated without the
publisher's prior consent in any form of binding or cover other than
that in which it is published and without a similar condition
including this condition being imposed on the subsequent purchaser

Contents

Preface

For some years I have conducted report-writing classes, mainly for scientists and engineers. When I was asked to write a book on the subject, I consulted a friend upon what form he thought it should take. His advice was to describe the sort of things I do in these classes. This is what I have tried to do. I assume the reader, as I do the student in the class, to have some knowledge of English, and so there are very few rules of grammar or punctuation to be found.

As most of the difficulties which crop up in reports are to do with writing rather than structuring the material, or wondering whether or not to capitalize a word, my practice is to put up on the blackboard those passages in the reports which I do not understand. Together we discuss the reasons for failure to communicate and attempt to categorize the faults. Hence the numerous examples in this book of selected extracts from typical reports.

Writing a book is different from taking a class, and in doing so I have drawn, in some cases overdrawn, from the experience of other people working in this field. My biggest debt is to B. C. Brookes of University College, London, who has not only allowed me to use much of his own material but has helped to structure my own thoughts upon the subject of technical writing. He and Professor R. O. Kapp, through their work with the Presentation of Technical Information Group, have done more than anyone in this country to popularize the importance of good technical writing.

I should also like to thank J. C. Y. Baker for filling a gap I could not have filled so expertly, in his chapter on 'Technical Illustration'; G. H. Wright for his appendix on 'Sources of Information'; M. W. Ivens, formerly Communication Manager for Esso Petroleum, for the guidance he gave me whilst working with that Company; R. W. Lewis, my colleague at Hatfield College of Technology, for the help he gave me both in the initial stages of writing this book and in suggesting improvements to the rough draft; Z. M. T. Tarkowski for his very useful distinction between Jargon and Technical Terminology, made in a paper to the P.T.I. Group at University College, London; and lastly the de Havilland Aircraft Co. Ltd for permission to quote so freely from their reports.

During my period with Esso Petroleum I was impressed by the way report writing was treated as a real industrial problem. Many courses on the subject were run by the company aimed at improving communication. Michael Hall conducted these classes with consummate skill. Some of the ideas he put across I have felt worth while including in this book.

Preface to Third Impression

It is encouraging that two reprints should have been called for so soon after initial publication. As the book has been so well received, I have limited myself to making only essential correction and alterations to the text.

I am grateful to friends (and reviewers) who have taken the trouble to point out inconsistencies and to make suggestions for improvement.

Acknowledgements

For permission to use copyright extracts the author and publishers offer their thanks to the following:

Messrs George Allen and Unwin Ltd, for an extract from *Technical Literature* by G. E. Williams.

Business Publications Limited in association with Messrs B. T. Batsford for the extracts and diagram from *Communication in Industry*, edited by Cecil Chisholm.

The Clarendon Press, Oxford for the extracts from H. W. and F. G. Fowler's *The King's English* and H. A. Treble and G. H. Vallins's *An A.B.C. of English Usage*.

The Director of Publications of H.M. Stationery Office for permission to reproduce the extract from *The Complete Plain Words* by Sir Ernest Gowers.

Messrs McGraw-Hill Book Company and Robert Gunning for permission to reproduce the Fog Index from the author's book, *The Technique of Clear Writing*.

Messrs Frederick Muller Ltd, for an extract from *Jet: The Story of a Pioneer*, by Sir Frank Whittle.

The *Spectator* for the extracts from articles by Bernard Levin.

The Times Educational Supplement and Professor C. S. Lewis for permission to reproduce a letter published in that journal.

The Plastics Division of Imperial Chemical Industries Ltd for the extracts reproduced from their *Technical Training Syllabus*.

Slip Products Co. Ltd, for two extracts from their pamphlet, *The Fight Against Friction*.

Vickers Ltd for permission to reproduce a page from one of their annual reports.

London Transport Executive for permission to reproduce the revised notices displayed in the buses.

Introduction

The purpose of this book is to examine how technical information may be written and presented so that it is easy to read and understand. It is addressed to the many scientists and engineers whose jobs require them to do more report writing than they would commonly like.

Is there a need for such a book? Do scientists and engineers do all that amount of writing to justify a study of it? Professor Edgeworth Johnstone, Lady Trent Professor of Chemical Engineering at Nottingham University, conducted a survey among practising chemical engineers to find out what they actually did in their jobs. The survey revealed that they spent almost one third of their working time on report writing, economics, and management.

How well do they perform this job of writing? If one is to judge from the amount of quicker reading and report-writing courses being held up and down the country, or from the remarks of senior management, the answer must be 'none too well'.

The reason for this deficiency is complex. In some cases it is the sheer inability to write clearly. The passage below was written by a very able and intelligent graduate engineer. It appeared after the title page of the report and thus was the first indication to the reader of what the report was about. It was entitled, 'Summary'.

The main theme of this Report is a justification for the elimination of pressure testing all copper tubing used for condensers and heat exchangers.

It is apparent that this testing has been carried out for a considerable period because of troubles experienced a number of years ago. Analysis of recent records, however, suggests that modern manufacturing techniques have improved the quality of materials to the point where reliance may be placed on the products of reputable suppliers.

If the main recommendations of this Report are accepted, the way is open for the complete re-organization of the Test Shop, integrating the Annealing Shop into it with resulting improved operator utilization.

Further method improvements following a re-layout of the Test Shop (made possible by released area no longer required for testing) will result in a final savings figure of approximately £20,000 per annum.

This passage is not untypical of that to be found in many reports. It is not all that difficult to understand what the writer is describing, as is the case in some reports. But if the purpose of writing is to reveal its meaning immediately and with the minimum of difficulty, this passage fails. It typifies many of the common faults which are to be found in reports, the avoidance of the first person and the use of the passive voice, 'where reliance may be placed' rather than 'where we may rely'. It shows a preference for the abstract word and for the cliché. The opening sentence would sound, as well as read, better if one said:

The main purpose of this report is to justify the elimination of pressure testing of all the copper tubing which is used for condensers and heat exchangers.

And what does 'improved operator utilization' mean? Does it mean that the existing operatives will be able to do more work, increase their output, do different and additional work, or does it mean that the company will be able to reduce the number of operatives? The expression has become a cliché.

Too many of the words carry too little meaning. What is 'a considerable period of time' and what are 'a number of years ago'? Three years? Six years? And what were the 'troubles experienced'? Fracturing of the tube? Shop-floor difficulties? What do expressions such as 'It is apparent that' and 'to the point' convey other than padding? 'Further method improvement' presumably means further improvements of method, but could mean further methods of improvement. Does a 'final savings figure of approximately £20,000 per annum' mean more than a 'savings of approximately £20,000 per annum'?

Could not the third paragraph have possibly been placed

second? Would it not have read better if a full stop had been placed after the phrase, 'integrating the Annealing Shop into it'?

Is this being pedantic? The plain fact is that you communicate more effectively when you adopt the active voice, when you write in shorter rather than longer sentences.

Communication is a more difficult act than is realized. It is strange that the craft of writing, the training in its skills, should for all intents and purposes end at the age of about sixteen, when most apprenticeships, whether they be in art or music, engineering or building, are just beginning. With the attainment of the General Certificate of Education in English Language at Ordinary Level, correction of writing ceases. Correction which has taken place earlier too often centres on such irrelevant rules as not starting sentences with the conjunctions 'but' and 'and', or the importance of avoiding sentences ending with prepositions. Sir Winston Churchill has nailed that one for all time with his classic comment, 'this is the sort of nonsense up with which I will not put'.

The observation of such rules as those mentioned above would not help the aeronautical engineer of a large and famous aircraft firm who writes in this manner:

However, in connexion with the canopy/fuselage fit, it has been disclosed that when the w/screen casting is received as a separate unit to the rear framework, the juncture of the two parts on each side often forms a step in the face mating with the woodwork. Further work is therefore needed to 'blend' the joint, before fitting and the situation is mentioned here so that consideration may be given to the question of the service being confronted with such a predicament.

Nor would blind rules have helped this chemical engineer:

The third digester batch in No. 7 digester was then got ready for pumping in the early hours of 25 March, but when the run-off valve was opened, no liquor was obtained. After stripping down the run-off system it was found that the run-off branch was choked completely solid with half-digested mass which no amount of rodding or steaming would clear. It was then decided to cut the back line and run-off by hose through this, then, having emptied the digester, alter the run-off system to common the back line and

run-off line upstream of the back line valve. This is a variation from
the West digesters where the back line valve and run-off valves are
on the same T-piece but the open end is to the digester base and not
the back line, but was thought to be expedient as it would give a
means of clearing the run-off line if it were to choke due to liquor
backing up during reduction.

Reasons for poor report writing

Why do people write in such a manner? I have suggested that
English training ends too early. That is only one factor. The
engineer from the aircraft firm does not write like this when
he writes to his wife or his parents or his friends. He may
defend technical writing of this order on the grounds of the
complexity of subject matter, but it is also a jargon which is
assumed for 'official' occasions. Its counterpart is still to be
found in what is mistakenly known as 'Commercial English',
with its letters to hand with thanks, your goodselves, and
similar twaddle. There is only one English and that is good
clear English. Another reason for the adoption of language
such as that in the example of the aeronautical engineer is the
curious idea that objectivity can be only established by writing
impersonally, that any intrusion of the observer somehow
invalidates his findings.

Not infrequently the writer adopts vagueness in the belief
that he is being diplomatic. No one can pin him down if he is
oblique. Equally he may play it safe and indulge in euphemism
and circumlocution so as not to hurt feelings or incriminate
other people. Truth in fact drains out of many a report as it
works its way up the management ladder. The net result is bad
English.

The aeronautical engineer has written as he has partly because he has rarely thought of an audience, so preoccupied has
he been with wrestling with his subject, and partly because he
has never really thought of the impact of words. What image do
the words and phrases – 'situation', 'consideration', 'confronted with such predicament' – conjure up in the mind? The
chemical engineer was cramming too many actions into one

sentence and using specialist jargon. The difficulty of writing
reports, therefore, arises from a failure to understand the
nature of communication itself and a failure to come to grips
with the human and organizational problems to be found
in most firms.

The nature of communication

To effect real communication means thinking much more of the
reader and much less about personal satisfaction. Transmission
is an easy thing to effect. The B.B.C. does this on the largest
scale every day of the year. But how much does it communicate? Those of you who use another channel will say 'not at
all': and even those of you who remain faithful throughout
the day will record varying degrees of reception. Success of
communication depends ultimately on the reader's or listener's
response. Genuine communication can only take place on the
basis of common experience and knowledge.

If you disregard written communication for the moment, you
will appreciate that communication can take place at a variety
of levels and in different ways. Obviously the dark auditorium
and the brilliantly spotlit arena, the gleaming black shirts
and breeches of the Nazis, the stirring Wagnerian music,
beamed out powerful messages before Hitler even opened his
mouth. So by our hair style, our clothes, and our accent, we
too, without being conscious of it, can produce attitudes of
approval and disapproval in our audience, which will effect
the reception of what we have to say.

We can use a variety of media to convey our information.
Our government publishes innumerable White Papers. Dr Fidel
Castro prefers to rely on television, a highly sensible method
to adopt for a largely illiterate people. In the firm how do you
go about conveying information? By telephone? Over the office
desk? By a notice on a board? By a memorandum? How often
do you think of the nature of your information and how appropriate are the means you use? A perusal of many notice-boards
would suggest that there are still many lessons to learn. The
Trade Unions, to judge from the drabness of some of the journals
they produce, seem typographically to be still living in the

nineteenth century. They appear not to have cast an eye on the
changes in lay-out which characterize most commercial publications. Because of the persuasive influence of television and
the cinema, in written publications there is a more marked preference for pictorial illustration. (Some firms, incidentally, employ closed-circuit television as a means of communicating
within the firm.) The Labour Party appreciated the importance
of varying its method of presenting its message when in 1958 it
engaged the *Daily Mirror* – the paper with the largest circulation in the country and brilliantly readable – to produce its first
glossy brochure, 'Your Personal Guide to the Future Labour
offers You'.

You should be aware how influential newspapers can be in
setting standards of presentation. 'Avoid long paragraphs. Solid
blocks of print weary the reader's eye.' So reads the advice
offered by one of the Schools of Journalism. Even the quality
papers observe this rule much more closely than they did in the
last century. Thus a single paragraph in a report – and I have
seen a fair number – extending over a whole length of foolscap
presents to the reader an unfamiliar experience. He is accus-
tomed to wide margins, double spacing, in most of his reading.
Why put up with anything less in a report? Likewise in the interests of lay-out, artistry, and appearance, there is a tendency
to dispense with punctuation marks that are conventional
rather than necessary for clarity of meaning. In a recent Annual
Report issued to the shareholders of Vickers Limited the following punctuation was observed:

LT-GENERAL THE LORD WEEKS KCB CBE DSO MC TD
SIR JAMES REID YOUNG CA FCIS
Barclays Bank Limited 54 Lombard Street, London EC 3

Who would say that the Board of Vickers Limited is illiterate
and does not know how to punctuate?

A person responsible for drafting, for example, safety regulations for laboratory assistants, the majority of whom are
secondary-modern leavers, could do a lot worse than take the
Daily Mirror for a week to see how that paper communicates
with its readers.

The written record, that adopted in reports, is the most

fallible method of conveying information. There is no feed-back taking place whilst the message is being studied. The diffi-culty is that the meaning of words varies with individuals. No word can evoke exactly the same response from two people. Similarly the interpretation given to two passages of writing will differ. Human beings are not logical mechanisms into which information can be fed. The practical implications of this are complicated when reports are intended, as they usually are, for more than one reader.

You should realize that there is no absolute relationship between either the sight or sound of words and their meaning. If I take two words, such as 'prolix' and 'furtix', unless you have seen them before and used them, you will be unable to unravel their meaning merely by studying them. The words have no meaning or significance in themselves. They derive meaning from what they refer to. If your past experience can provide no reference there will be no meaning. The following two diagrams help to illustrate the point. Figure 1 shows how communication takes place:

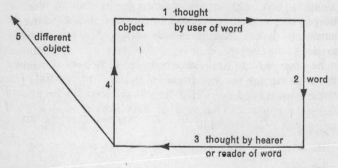

Figure 1

In Figure 1 you will observe that there is no diagonal line linking 'word' to 'object'.

Figure 2 (overleaf) attempts to represent the shared and separate knowledge of two specialists.

If the average educated man's vocabulary is about 15,000

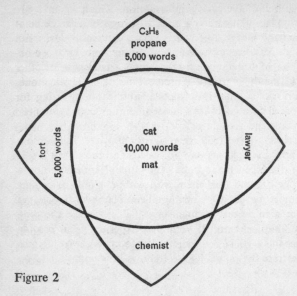

Figure 2

words, 10,000 of those are common to the chemist and the lawyer. Words such as propane, or formulae such as C_3H_8, are particular to the chemist; words such as tort are familiar mainly to the lawyer.

By now you will have consulted your dictionary and found the meaning for the word 'prolix' but not for 'furtix'. The latter word is one I fabricated. It looks and sounds all right and serves to emphasize that you cannot discover what a word means just by looking at it.

Writing within the firm

I have considered just a few of the problems inherent in communication. Many difficulties arise from the organizational problems posed by the firm. How many types of management are there? What are the written and unwritten conventions which govern writing reports? Who takes the decisions? What is past practice?

What, for instance, is the climate in which reports are written? Report writing is an activity which is often done at home, partly because a person writing a report does not appear to be working as hard as someone employed more obviously with a drawing board or some apparatus. Another reason, of course, is that the office is noisy with telephones ringing and other interruptions, so that the ideal setting for composition is not easy to find. How encouraging and beneficial it would be if firms could provide a few soundproof cubicles grouped round a small reference library.

LACK OF ACKNOWLEDGEMENT Again, a factor which militates against good report writing is the lack of acknowledgement which a report frequently receives. Work which is appreciated induces care of composition by the author. Report writers do not expect bouquets, but an occasional remark or note of congratulation would not come amiss. The circulation list on some reports is so large that probably no one regards it as his responsibility to encourage the author. The result is that report writing can too easily become a mere chore, and under such conditions it is seldom done well.

OVER-EDITING Sometimes the opposite process takes place where a report is subjected to the most detailed departmental editing. Perhaps an unusual example was the case where an engineer received his report back from his section head for re-drafting and re-submission to a departmental head. The departmental head returned it to the author with his own comments for re-drafting and re-submission, this time to a group of three departmental heads, who in their turn sent it back to the author for a third re-drafting for final submission to the ultimate reader. Obviously a manager has a right and responsibility to see that his subordinates do their job properly and so reflect credit on the department. Very often this criticism centres round trivia, a dislike of split infinitives, a certain type of abbreviation or punctuation. At its lowest level this criticism merely affords an opportunity for the exercise of authority.

VAGUENESS OF SPECIFICATION Very many reports are written badly because the person who called for the report initially gave imprecise instructions. Because his original specification was vague, the resulting report does not hit the target but covers a lot of irrelevant ground. This is very common indeed. The author imperfectly understands what he is required to do, and for fear of being regarded as dense, composes what he thinks are the originator's intentions rather than return and ask for something more explicit. You should seek elucidation at the risk of minor annoyance rather than press on in the dark and much later cause even more aggravation by a bad report, which suggests amongst other things that you have been wasting your time. For your boss will probably send you away on a report-writing course with the instruction 'teach this man *how* to write' when what you really need to be told is '*what* to write'. One might add also that you usually write better when you know *why* you are writing. What? why? and for whom? Know the answers to these questions and your chances of writing a good report are improved.

The importance of specification is nowhere more necessary than in the matter of building a house. There is a twofold breakdown in communication. The client never conveys accurately or technically enough to the architect what it is he wants. The architect's instructions to the builder also seem to be full of occasions for misunderstanding.

What is style?

Bad reports frequently result from joint or multiple authorship. The most successful report is when one person has given the report a unity by a single style. He may have had to assimilate facts and opinions from a great many people. He may have done little of the groundwork. If he merely re-arranges sections written by others and interlards sections of his own, the overall effect is one of neutrality. Far better to rewrite the lot.

I have used the word 'style' in the paragraph above. It is the word from which many writers, particularly scientific ones, will retreat, as they mistakenly identify it with something varnished, a lacquer added on afterwards to the bald narrative in order to achieve a pleasing finish. Style is how you or I write, not just how some classic writer wrote or how some modern novelist writes. People will write differently and better than others according to the range of their vocabulary, their sensitivity to the rhythms present in language, and the structure and balance of a sentence. Certainly some passages rich in metaphorical language may be described – though somewhat inadequately – as flowery style. But a plain, terse, straightforward account of a piece of research also possesses a style which is recognizable and, one hopes, pleasing. Thus, if the most effective communication is to be found between two individuals, this is a strong argument for single authorship, even though the work described in the report may have engaged the attention of many. It is easier to limit the authorship to one than the readership.

This whole question of style is one which will be taken up at length in another chapter. It is crucial to the whole problem of report writing. So often you come across reports which are difficult and boring to read because the writer takes no trouble to make his prose various and interesting. He adopts the impersonal style, with all its disadvantages, as an assurance of his own objectivity, or as a safeguard against being tied down too closely to what he has said.

If there is one criterion you should have before you in a world where almost too much is written, it is to think often of your readers. This is easier if you ask yourself, 'What do I enjoy reading?' You are sure to say publications which are nicely printed, well laid out, with good spacing, convenient to handle, and easy and pleasurable to read.

The remainder of the book will discuss ways in which these effects can be produced.

NOTE: Those interested in the theory of communication, namely the use of symbols and signs by which men influence each other, are thoroughly recommended to read *Studies in Communication* (Secker and Warburg, 1955). These are a series of papers contributed, mainly by university staff, to the Communication Research Centre at University College, London. The papers range widely and include such topics as the study of machines that transmit, process, and store information, the social factors involved in communication, and the biological study of how an efficient memory store can then be built up. There are essays contributed by a phonetician on speech mechanisms, by a Professor of Fine Art on the interpretation of visual symbols, and by a Professor of Language on the influence of language on medicine.

As a report writer you should consider the social factors involved in communication. They involve such matters as the size of the group, whether the unit of social studies is industrial, military, governmental, or nationalized.

Professor B. Evans who edited the book stressed the problem of how the semantic content of a message changes as it passes from one level to another, as for instance from 'top management' in an institution down to 'working level'.

1. What is a report?

In the Introduction I have dealt with some of the difficulties inherent in communication. I have suggested that these difficulties can be aggravated by organizational and human relations problems. Many of the points I have made affect all types of communication so it is as well to ask, 'Is a report a particular kind of communication, well differentiated from all other kinds?'

What is a report?

There is no absolutely satisfactory answer to this question. In a very general sense a report is an account of something. We talk of school reports, of newspaper reports, of Parliamentary Reports, of Law Reports, to mention but a few types. Have these anything in common? Of the four listed above, three follow a certain standardized form (newspaper reports can take almost any form). If we feel our way towards a definition of a report, therefore, 'form' would be one of its characteristics.

In essence though, what distinguishes a report – particularly a technical report – from, say, a novel is that a report conveys certain specific information to a specific reader or readers. It is not written because you happen to feel good one morning and want to reveal to mankind your particular joys. A report is usually an answer to a question, or a demand from some other person for information. 'I want to know why this metal is fracturing', 'What are the chances of improving our exports to Sweden?', 'Get me a statement on the number of employees using the pension scheme'.

You need not convey all requests for information by means of a report. You can use a letter or a memorandum or a note. How does a report differ from these other methods of communication? By the presence of a certain 'form'? You can get confused here by the difficulties of nomenclature. If you were asked to produce a note or memorandum on the uses which had been made of a recently acquired computer in your firm, would what you produce look very different from a *report* on the same subject?

The question is only raised in order to underline the fact that firms do use different methods of communicating and employ different terms. Sometimes it is more appropriate to convey your information by letter and sometimes by report. As a rough guide reports will be preferred to letters and memoranda when the information to be conveyed is longer and when it is addressed to a number of readers. This is only a very rough guide and firms differ in their practice. Letters do tend to be more personal documents. They are written on flimsier paper and rarely have a protective folder, which makes for ease and frequency of handling.

It is not true to say that reports have a structure and letters do not. A lengthy letter (or even a short one, for that matter) will have a main heading, which in reports we call titles; it will have capitalized and underlined headings and sub-headings; it will, one hopes, possess logical development. Its mode of address will, of course, be different.

In practice you have a reasonable idea of what a report is and looks like. Many of the large firms have handbooks outlining their own procedure. The newcomer finds out about writing reports from this or by familiarizing himself with previous reports. This is not always a good thing. Sometimes it would be better if he started out in the void and was forced to ask himself what was the nature of the matter he had to convey and what was the best method of conveying it.

Classification of reports

Attempts are often made to classify reports. Company handbooks are not such offenders as some textbooks on report writing, which are notorious in their classification of the different types of reports. I have seen the following types of reports listed in a number of textbooks: Preliminary Reports, Interim Reports, Inspection Reports, Decision Reports, Operation Reports, Construction Reports, Valuation Reports, Design Reports, Investigation Reports, Test Reports, Formal Reports, and Informal Reports. The danger of this categorization – much commoner in American books on the subject – is that it suggests that these categories are differentiated types each with their own characteristic treatment and lay-out. This may be so for a few. Companies may print Test or Inspection Reports for use with routine checks. Generally, however, the use of such terms can be misleading. You are not asked to produce, say, a Design Report or an Operational Report in the same way that you are asked to go to the stores and collect a definite size of spanner.

I will consider this matter of lay-out in a later chapter, but the nature of a report and its purpose will dictate its structure. It is dangerous to lay down formulae for certain types of report. This only encourages you to squeeze your material into some set mould rather than allow the material to find its own shape.

Some categorization, such as that practised by a well-known engineering company, is sensible and useful. The categories are related to usage and readership. Category 2 is Non-confidential. Reports under this heading have a wide circulation within the company and collaborating companies. Category 3 is Confidential, which is what it implies, and reports are circulated only to people with a direct and immediate need for it. Category 4 is Secret with very restricted readership which has to be authorized at high level. Category 5 is Internal. Reports under this heading are of purely local interest.

If you think less in terms of categories and more in terms of what you are being asked to do, you will arrive at a better answer. Is the Report you have been asked to write a request for information? If so, your task should be a brief and lucid statement of the facts. If on the other hand you are answering a request for a report on losses, some explanation is necessary, some justification possibly. Is there an implied or real criticism in the request? And what approach should you make by way of answer? Are you going to be honest?

Or you may be asked to recommend a certain course of action. This would be a case where persuasive language is called for, a skilful deployment of arguments for and against, and where it is important to know something of your readers' attitudes and feelings. This line of thinking is more profitable than asking yourself: 'Have I an inspection report to write? What is the form for that?'

Writing for a varied readership

Although someone in authority initiates a report, you will not only have to cater for the initiator, but in all probability a large number of other people too. Some of these people will only read the report cursorily, others will have to scrutinize it minutely. The difficulty is in satisfying all readers in their own way. To communicate most effectively therefore you must know the readers' knowledge, needs, and attitudes. Too much detail will bore some readers but appear inadequate to others. And if writing is a matter of tone and intention as well as the mere conveying of information, it is important to know what are the readers' feelings about the subject and therefore what is the most tactful approach to adopt.

The problems of catering for a mixed readership will be discussed in a later chapter. The use which readers have for a report will determine what selection of material is made, what method of presentation it is most logical to follow, what amount of interpretation is necessary. The important thing always is to keep the reader firmly in mind. A person responsible for drafting safety regulations for perusal by secondary-modern leavers, who is himself normally passing on highly technical research work to well-qualified fellow scientists, will have to make some adjustment in approach and style.

Why reports are written

We might ask ourselves why reports are written. In the old days, of course, they were not. When industrial units were much smaller and the industrial processes less complex than they are today, it was enough for the manager to walk down to the shop floor or laboratory and obtain a verbal report. The larger and more dispersed an organization becomes, the more dependent it becomes on documents.

Reports will serve as evidence of work that has been done. They are at the same time the essential records of that work. They are valuable as a repository of information, possibly of great capital value to a company.

At the same time they may serve as the basis for decision and action. So many factors have to be taken into consideration nowadays before a decision is made that it is necessary to have all the relevant information in written form. The great advantage of reports over oral communication is that they afford the reader the opportunity to study the contents at his leisure or at a suitable moment. It also means that many people can read them at the same time.

Although reports are not written to enhance the prestige of a company they do nevertheless serve this purpose. This is particularly so with research departments. The quality and quantity of their scientific work is judged by the quality and quantity of their published papers.

Reports serve the not altogether unworthy aim of providing a challenge to the individual to be explicit and articulate. This is the crucial test of a person's knowledge, whether he can make it intelligible to others. This is surely the difference between the technologist and technician, that the former is able to write and talk about his job in abstract, general terms. Science and technology are essentially public activities.

Alas, too often the scientist or engineer looks upon report writing as the least important and necessary of his activities, but science begins only when the worker has recorded his results and conclusions in terms intelligible to at least one other person qualified to dispute them. In fact, the scientist or engineer tends to separate his practical work of observation and computation from the writing of his account of it. Ideally the activities should be considered together, his writing the natural and logical expression of his thought.

2. Preparing the first draft of the report

I have suggested that it is better to regard report writing as the conveying of certain specific information to a specific readership. Although a report does have a certain form which most of us can recognize, there is no absolutely agreed way of laying out a report or even of determining at what stage the material required should go in letter form or in report form. Even less is there common acceptance of what constitutes the difference between a Valuation Report or a Progress Report.

It is more profitable therefore to consider the total situation in which the report is written than to be tied down with categories of reports. The main elements (in the total situation) are the writer, the reader, and the material, or the subject that is being written about. One could add the purpose of the writing and the circumstances that provide the occasion for the writing. If during the writing of the report you return to these fundamentals you are less likely to go wrong than wondering whether you are observing the canons of 'correct' report writing.

The main stages in writing the first draft

The main stages in writing the first draft of the report – and it is usually necessary to write a first draft – are: (a) Collection of material, (b) Selection, (c) Logical ordering, (d) Interpretation, (e) Presentation. At all these stages you should think of the main elements, namely, the writer, the reader, and the material. Many writers will concentrate on one element to the exclusion of the others. The easiest element to concentrate on is the material or subject. A typical example is where a technical man is writing a report advising non-technical people about the advantages of purchasing and installing a new technical device. As Professor R. O. Kapp says:

The mathematical analysis that enables the manufacturer of the device to determine the leading parameters would not be relevant to the whole situation. Yet this is just the sort of thing that too many writers seem to feel must be included in their reports willy nilly; and it shows that they are thinking about the subject only, and are ignoring important aspects of the situation as a whole. They would avoid the error if they asked themselves the question: What is going to be done about this mathematical analysis if I include it? Is it helpful to those who will have to reach a decision? Suppose, further, that the device is suitable only for operation from low-impedance sources. The information is certainly relevant to the *subject*; but it would be relevant to the *situation* only if the readers knew what a low-impedance source was. If they would not recognize a low-impedance source when they saw one, they would need something more than that mystifying piece of information. A writer who appreciated the total situation would therefore elaborate with such additional information about low-impedance sources as would make them recognizable to the reader.

You should, then, ask yourself, not only at the beginning of writing a report, but also throughout the report, 'Why is this being written? What ought to be done about it? If I want action, are my recommendations clear enough, specific enough? Have I defined the expected result quite precisely to myself?'

Obviously your answer to 'Why is this being written?' is most likely to be because someone told you to write it. At the back of your mind therefore should be the image of the reader or readers. Remember that the report is not being written for your benefit but for use by someone else. It is not important what you think of the report, how it satisfies you, but how it satisfies somebody else, possibly very different from you.

TYPES OF READERS What types of readers are there? There are the administrators and executives, who may or may not have been specialists. There are the specialist science readers and the general science readers. There are supervisors, colleagues, and subordinates. Some of these categories overlap. You should consider your relationship with your reader, for this will effect your tone. If the report is a persuasive report in which you are recommending a course of action, your reader's *feelings* are important. What reaction do you want to arouse?

Having got some picture of your readers and the purpose and scope of the report, you can start on the first stage of the report – the collection of the material.

Collection of the material

There is no standard way of collecting material. How you go about it will depend very much on the circumstances. If the workers in a particular shop do not return to work after their lunch break, and the managing director asks the labour-relations officer to let him have a report on his desk by nine o'clock the following morning on the nature of the dispute, it does not leave much time for trotting along to the local reference library to build up the background of trade-union history. On the other hand the report writer may be given three months' grace to report on research developments in the United States.

This is only to say that sometimes you will use reference books and textbooks, sometimes periodicals and other reports, sometimes catalogues and patent specifications. At other times you will rely on your own observations. These may be as a result of experiments you have performed, or interviews you have conducted.

In the Appendix sources of information are described. You should certainly use some of the reference and advisory facilities mentioned. Human knowledge has become so vast that it is only with the help of such organizations as the Association of Special Libraries and Information Bureaux (ASLIB) that you can hope to find your way about. Particularly valuable is the abstract service provided by some technical colleges or organized by the Institution of Electrical Engineers in their *Science Abstracts* or to be found in such publications as *Engineer's Digest*.

The important rules to remember when collecting information are: (a) to check the accuracy of your facts; (b) to separate facts from opinions – and assess the merits of the opinions; (c) to separate facts from inferences – and see if the inferences are well-based.

A firm may be having trouble with the fracturing of certain

components. If you are asked to investigate the cause of this and the foreman in charge of the shop tells you it is because the machine is too old, or the temperature too high, or the pressure too low, it is well to remember, unless you have made a thorough scientific investigation of the matter, that it is the foreman's *opinion* that you are recording. If he has been with the company thirty years and is the most experienced and knowledgeable man in the shop, his opinion is obviously worth a great deal.

Good examples of opinions presented as facts can be found in almost any foreign correspondent's dispatch, where even mere café gossip will be phrased as though it had come straight from official government spokesmen.

The golden rule to keep in mind when collecting information is a truism – namely, collect only relevant information. You are thrown back again on our main elements (the writer, the reader, the material). What is *relevant* to your particular readers?

Selection of material

The processes of collection and selection of material are not isolated but obviously overlap. You must be making some selection whilst you are collecting material. It would be ludicrously wasteful to look up all you could on a subject and then make a selection of it to suit your reader's requirements. On looking over your material, you may find on reflection that you have still got too much – or equally that you have left an important item out.

CLASSIFICATION OF MATERIAL It is difficult to know whether the classification of your material should more appropriately be treated under the heading of 'Collection of Material' or 'Selection of Material'.

If your report is a long one involving much research and preliminary inquiry you cannot hope to record all your information on the back of a writing pad. If your inquiries extend over some months your memory may fail you and you will want some method of recording which allows for easy reference. It is not necessary at this stage, nor even likely, that you will classify your information in the manner in which your report will ultimately be presented. The headings will inevitably be broad ones, such as 'History', 'Equipment', 'Background', 'Research Work'. These will admit of sub-headings, so that the last one might be divided into 'Research Findings in America', 'Research Findings in Germany', with further sub-headings for products, if more than one is being considered, or for teams of workers, if there are many in the field.

Your original specification may have given you some guide as to the most appropriate headings under which to classify your work. If you were writing a paper on the causes of soil erosion, it could be classified in two main divisions, 'Erosion due to Nature' and 'Erosion due to Human Intervention'. Under the first heading your sub-headings might well be 'Erosion by Water' (or lack of it), 'Erosion by Wind', 'Erosion by Fire', or by a combination of two of them. Under the second heading your sub-headings might be 'Erosion as a Result of Accident', 'Erosion as a Result of Abuse'. The first example might be a more appropriate arrangement for an article in a learned magazine; the second example a more appropriate approach for a report to the Forestry Commission or some public body who would wish to take immediate action as a result of the report. They would be familiar already with the classic reasons for soil erosion and doing their best to cope at this level.

You might be asked to report on the production methods or management training employed by another firm. Your division for the latter could be, one, 'Types of Training', with sub-headings for 'Works Study', 'Human Relations', 'Communication Theory', 'Sales Techniques', and, two, 'Levels of Training' with sub-headings for 'Executive', 'Supervisory', or 'Operator' level.

A report on the performance of an aircraft would normally break down into such headings as 'Flying Controls', 'Systems', 'Flight Characteristics', 'Special Equipment', 'First Flight Data'. These headings would have appropriate sub-headings: for 'Flying Controls' – 'Elevators', 'Airlerons', 'Rudder',

'Flaps', 'Dive Brakes'; for 'Systems' – 'Fuel Tanks', 'Air Conditioning', 'Water Ballast'. And so on.

FILING INFORMATION You can file your information in a number of ways. The best method is to use some loose-leaf folder or card cabinet. This enables you to place material from different sources together, for example, a graph alongside some written material obtained elsewhere. It also allows you to add material at a later date without the embarrassment and untidiness of writing between paragraphs already written. With a complex report, some system of numbering or coding may be necessary or the use of coloured tag-strips for cross-referencing. A very detailed report might require a decimal system of classification or even the use of a punched-card system.

How you actually record your evidence is a personal matter. You may have some elaborate system of abbreviations, known only to yourself. This is all right provided no one else has to work on your material. You may have a good memory and need only short notes rather than lengthy extracts. In the university or college notebooks kept by students there is no correlation between the amount of notes taken and the class of honours degree obtained.

The whole approach to classification and note-taking will vary according to the individual and the terms of reference. Given some choice, one person will prefer the personal interviews whilst another will be content to sit on his bottom and do all the work from a desk hid beneath a mountain of books. The approach to collecting and classifying information is not in itself important. What does matter is: does the system work well and efficiently for you? If your own methods seem laborious then it is time to adopt some other method, a decimal or punched-card system.

IMPORTANCE OF ACCURACY You should note fully the source of your information. There is the danger of recording a fact as being a vital piece of information without taking precise details of your source. Where possible, record on the spot, even

if this means a double recording. Use the back of an envelope
if necessary. Original impressions soon fade. Diagrams and
sketches, though perhaps not essential to the work at hand, can
be a useful *aide-mémoire*.

THE ACTUAL SELECTION When you have collected and classified your material, you must make some selection for your first
draft. If your initial planning and thinking about the report
have been thorough, and if new circumstances have not arisen
to invalidate your findings, there should not be much material to
reject. It should be more a matter of arrangement, of isolating
the important from the unimportant, of obtaining the right
degree of emphasis and subordination.

The act of selection – much more so than the initial investigation, where the subject matter seems most important – should
be concerned with the reader. This is a platitude but is often
overlooked. You should be aware of the extent and nature of
the knowledge of your readers, their interests, their relationship
to you, their status and authority, and even your own authority.
These will have some bearing on the choice of material, the
sequence in which the items are presented, the method of
presentation, the level at which the matter is discussed.

You should leave out items which are beyond the understanding of your readers, items which are too well-known, items
upon which you cannot speak with much authority, and items
which will not help towards a decision on action.

Logical ordering

Problems arising from selection can be equally well discussed
under the heading of 'Logical Ordering'. Your method of investigation must inevitably be chronological. Even if you have
divided your work up into sections, you must start somewhere,
with one section, and then go on to study the others. What is
frequently the case is that this becomes the method of presentation in the report, in other words chronological ordering rather
than logical ordering. Very rarely is this appropriate. Usually

this method does not sufficiently highlight the important aspects of the report. There is no light and shade.

Some graduate trainees from a large chemical firm were asked to write a report on a three months' Production Engineering course they had attended at a technical college. The course covered such subjects as workshop practice, jig and tool design, work study, and embraced a project and a fair number of works visits. The report was not an untypical exercise set by firms which organize some of their workshop training outside their own firm. Few of the graduates whose reports I studied knew for whom they were writing. The education officer responsible for their training had told them he wanted a report on the course. The graduates assumed their report would be read by this education officer, but did not know if other people would read it. Worst of all they did not know what they had to write about. The specification was general rather than precise. Were they being asked to evaluate the post-graduate training they were receiving? Were they being asked to display their familiarity with the processes they were studying? Were they supposed to *impress* the education officer – and if so, how should they go about this?

The results, needless to say, were deplorable. Many of the graduates relied upon a purely chronological narration of their twelve weeks spent at the company's expense, namely, what they did on their first day, what they did for the first two weeks, how they did the project in the penultimate week, and so on. A few assumed they had been asked to evaluate the training received and divided their reports under such headings as 'Quality of Lecturing', or boldly started off with an account of the works visits and treated rather briefly of the routine work done in the workshop – presumably on the assumption that the reader would be more than familiar with the facts of how to make a weld with oxy-acetylene, and know rather less about the firms visited and the value of such visits.

This is to emphasize that the material can be presented variously, and that the rejection or selection of material, and the logical ordering, must be related to the reader. It is a fair rule to say that the most important parts of a report come either at the beginning or the very end, but preferably at the beginning.

A reader should be able to pick up a report and be able to say very quickly indeed what the report is about.

Another example of failure to select and order was the case of a fuel consultant organization. They issued a sixty-page report on the completion of their investigations. The report was addressed to the managing director. Included in the report were no less than forty-nine recommendations. Some of these were as trivial as the changing of a washer in a boiler house, others were drastic, involving the firm in complete renewal of plant and considerable sums of money. All these were solemnly recorded in what seemed chronological order, i.e. in the order in which the various defects had been discovered.

The beginning of the report was no help. This dealt with the background, the method of study adopted. The sort of statement that the managing director would have welcomed right at the very beginning would have been something to the effect that basically the fuel efficiency was good, or adequate, with the exception of some particular corner of the works, or, alternatively, that the place was likely to blow up any minute unless immediate steps were taken to repair or replace some power unit. The smaller items, which would more fittingly be addressed to the Maintenance Engineer, could have been recorded in an Appendix.

The following actual but disguised report, although it pretends to have some semblance of structure, is really a piece of chronological reporting:

Degradation of Quality on Stocks of Gasoline and Turbo Fuel at Bognor Installations

In recent months two instances of degradation of quality on stocks of gasoline and turbo fuel occurred at Bognor installation. To discuss this problem, Mr Smith of Engineering Department, Mr Jones of Technical Sales Department, and Mr Robinson and Mr Brown of Supply Department went to Bognor on 1 August. The problem was put to Mr White of Bognor, who suggested that the trouble was mainly due to 'weathering'. However, it was pointed out that comparable conditions applied at some other terminals with regard to the small throughputs, which led to considerable ullage existing in the tankage for several months before being closed by replenishments.

Bognor suggested that the tankage in question might be inspected and the following conditions were found to exist:

Tank 200:

This tank stores gasoline, has a capacity of 1,000 tons and, at the time of writing had a stock of approximately 400 tons. The Reid Vapour Pressure of this material had dropped to 5.1 p.s.i. against the specification minimum of 5.5 p.s.i. Mr White had previously pointed out that during June there was a loss of 10 tons due to wide diurnal fluctuations in June. Mr Jones and Mr Smith examined the roof of this tank, where they found vapour escaping from the side of the roof, a diphatch cover left open and there were no P and V valves fitted.

Tank 400:

This tank stores turbo fuel, has a capacity of 1,000 tons and, at the time of writing, had a stock of approximately 800 tons. This material had been transferred from Tank 600, which has a capacity of 5,000 tons and, after the transfer when testing was carried out, the R.V.P. was found to be 1.5 p.s.i. against the specification minimum of 2.0. The roof of this tank appeared to be in better condition than Tank 200 and did not appear to have P and V valves fitted. As in the case of Tank 200, the diphatch covers had been left open. In addition, it was reported that Bognor make a practice of casing their pipelines when transferring from one tank to another by 'air blowing' and this, of course, would have the tendency to carry the lighter fractions out with it.

In the past it has not been normal practice to test the before and after transfer of material from one tank to another, but it may be that when degradation of this sort takes place, that a sample prior to transfer would be useful in assessing whether degradation had taken place.

After leaving Bognor, it was agreed that Mr Smith of Engineering Department would submit a more detailed report for Technical Sales and Supply Departments to take the necessary action. It was suggested that Technical Sales would wish to act as regards the quality for A.I.D. control and Supply Department would act on the question of stock losses and general operations facilities at Bognor.

Supply Department	A. Robinson
2 August 1957	B. Brown
AR/MB	

You have to fish about in this report, read the matter more

than once, before you have an idea of what the writer is reporting about.

Very few people really enjoy reading reports; nobody enjoys reading them twice. If the writers of this report had asked themselves: how can we write an account of our work so that the reader will not have to read it twice? rather than: how shall we write up a morning's work? they would have produced something better than they have done.

The failure of this report is not only in the purely chronological reporting but in the failure to categorize and assess. Two of the team of four sent down submitted this report to a senior manager. It is not apparent till the very end that this is a Preliminary Report. This should have been made obvious from the start. The title might well have been changed to, 'Possible Causes of Degradation of Quality on Stocks of Gasoline and Turbo Fuel at Bognor Installations'.

The only categorization that has been made is the two different tanks that were inspected. Presumably Tank 200 comes before Tank 400 because 200 is a smaller number, or more likely because the investigators looked at Tank 200 first. Headings more appropriate to the title might have been 'Gasoline – Tank 200' and 'Turbo Fuel – Tank 400'. But even these would miss the point.

Can you pick out easily from the report what are the possible causes of degradation? Close analysis would reveal that there are three possible main causes of degradation, namely, weathering, faulty equipment, and malpractices.

An amended structure of the same material might read as follows.

Introduction   A statement that this was a preliminary report on the possible causes of degradation of quality on stocks of gasoline and turbo fuel at Bognor Installations. To be followed by fuller reports.

The background to the request for the investigation. Brief analysis of the likely causes.

Main Discussion of Report
  1. *Flaws in Equipment*
              Tank 200 – Amount of loss
                          (a) No P and V valves
                          (b) Leak in roof

> *Tank 400* – Amount of loss
> >   (a) No P and V valves

2. *Malpractices*
> >   (a) Diphatch covers left open
> >   (b) Air blowing
> >   (c) No sampling

3. *Weathering*

Conclusions and Recommendations
> >   Tentative conclusions pending more detailed re-
> >   port. Allocation of specific responsibilities.

The language in the original report could itself well be improved. What is one to make of a sentence such as 'The roof of
this tank appeared to be in better condition than Tank 200 and
did not appear to have P and V valves.' If the first 'appear' is
justifiable surely the second one is not. It is correct to say that
the roof appeared to be in better condition. (Closer examination may have revealed it not to be.) But the tank either had
P and V valves fitted or it did not.

What I have given is only a skeletal structure and could possibly be improved, but it does enable the reader to disentangle
the issues more easily. If action is to be taken it is important
that the issues should be laid bare in this way. If the final report
emphasized the malpractices as being primarily responsible,
this would suggest negligence on the part of the supervisor. If
the equipment was faulty, this might be the lethargy of the
divisional maintenance department. If weathering was accepted
this would exonerate both parties and throw responsibility on
the Almighty.

The logical ordering is that which tells the reader straight
away what are the likely causes for degradation of fuel, who
is responsible, who wants shaking up. The first report is likely
to smother action.

LEVELS OF READERSHIP If a report is to be read by a number of readers of differing degrees of authority and importance,
the logical ordering will be to present in a nutshell for the top
reader the necessary information as briefly and as early as
possible.

Theoretically it would be possible to have four different levels of readership. At the highest level, little more than the title and evidence that the work is to hand may be enough. At a lower level the manager will want to familiarize himself with the conclusions and recommendations. Further down someone will have to study the argument of the report in detail. At the lowest level of readership the footnotes, authorities, and appendices might have to be checked.

Professor R. O. Kapp presents an interesting example of a case in which there is some difficulty about ordering.

The general case is when a point, A, cannot be appreciated until another point, B, has been understood; and point B cannot be appreciated until point A has been understood.

Suppose, for instance, that a report contains a recommendation, the nature of which is not easy to understand, and the need for which is not at all apparent to the reader – the person who will have to act upon the report. (He may even have a strong objection to any change.) To make the nature of the recommendation clear, a number of technical points have to be developed at length; and the need for the action recommended can only become clear to the reader when he has understood these technical points. Now it may well be that the reader will not have the patience even to try to understand them (and consequently the nature of the recommendation) until he has become convinced that some action is necessary. There is the writer's dilemma. If the reasons for the action are presented first, they will be incomprehensible; if the recommendations are presented first, they will not be listened to. So this is one of the problems that a report writer is sometimes faced with – where the sequence in which items should be presented according to the logic of the subject conflicts with the sequence dictated by the logic of the whole situation.

METHOD OF ORDERING MATERIAL Remember, then, that there are a number of ways structuring your material. The chronological is only one of them, it is suitable at times, it describes the way you collect your material, it is very often the easiest method to adopt but not always the most appropriate.

Other ways which are open to you are the categorical method, the comparative method, the logical method, to name a few.

The categorical method would be to start with a general statement and work down to particular points. Computers might be the general class; analogue and digital the particular; mechanical, electro-mechanical, and electronics, the more particular; and so on.

The comparative method is usually combined with another method, such as the categorical. Employees are frequently asked to find out how a rival organization recruit their technical officers, what expenses they allow for travelling auditors, how their processes for the same product differ. It is convenient to categorize such information under suitable headings and to present it side by side with the policy of the employee's own company.

The logical method may be sequential. Sir Frank Whittle in his autobiography *Jet* describes the major organs of a turbojet engine. 'These are,' he says, 'a compressor, a combustion chamber assembly, a turbine, and an exhaust pipe ending in a jet-nozzle. Large quantities of air drawn in at a front intake pass through these organs in that order.'

A report on a jet engine might well have been structured in that way, logically, in terms of the action of the engine. Thus you would have four main headings in your report in the order above, 'Compressor', 'Combustion Chamber', 'Turbine', 'Exhaust Pipe and Jet Nozzle'.

Other methods of structuring could be based on order of importance or on cause and effect. Common sense will generally dictate the form which the report should take.

Interpretation

Interpretation is to do with the level of presentation and the level of language. You may have made the most effective selection of information, and marshalled your evidence in the most logical order, but you are still left with the difficulty of knowing at what level to pitch your language.

This is not just a matter of style, which will be discussed later, but of understanding how familiar your reader will be with your terminology. I discussed, earlier on, some of the difficulties inherent in any communication. We tend to assume too readily that people understand us. The little research that has been done on this makes depressing reading.

GETTING THROUGH TO THE AUDIENCE The B.B.C. Audience Research Department investigated listeners' reactions in 1951–2 to their highly successful and topical Light Programme talk *Topic for Tonight*. It is worth pointing out that this five-minute general talk is devised to interest the largest possible audience and is given by a member of a large panel of experts and journalists familiar with broadcasting. It has been immune from criticism. The B.B.C. had reason to feel proud of their efforts at popular enlightenment. The average score, however, in comprehensibility tests was twenty-eight per cent. An account of this in *Communication in Industry* says:

This means an understanding of 'intake' of about a quarter of the ideas and information given. Three-quarters of the material passed completely over the hearers' heads. Average score for individual talks varied widely from eleven to forty-four per cent. Poorly understood talks tended to involve a serious over-estimation of the listener's background knowledge. The speaker used too many difficult words. He made too little use of summaries or other methods for driving home his points.... Apparently other speakers lost their hearers' interest through 'talking down' to them. Yet these five-minute talks are exceptional by any test. They are heard by thirty per cent of the adult population during each week.

The same book has an interesting example of a survey done by 'Mass Observation' in 1947.

In 1947 the Government decided to put out a popular version of the annual *Economic Survey*. So they gave it a coloured cover and added diagrams to explain the figures in the text. This, they felt sure, would get the *Economic Survey* read. Mr Attlee, the Prime Minister, expressed his confidence of this in the Commons. A few days later Mass Observation asked a carefully chosen sample of the population how much of the pamphlet they had understood. They found that the whole thing, cover, diagrams and all was practically unintelligible to the lower middle and working classes, skilled and unskilled alike. None of the women, few of the men, were interested in the text. Of these only a very small number indeed could understand it sufficiently to grasp its main points. It takes more than a few coloured trimmings to make a statistical report interesting.

The British Institute of Management conducted a survey of annual reports where firms attempted to give financial information to their employees. The number that did so successfully was small.

You may say that not many who read this book will be required to prepare a financial report for ten thousand employees – although someone has to do it – and that your problems of communication are much simpler. A request of this sort does, however, highlight some of the difficulties involved in communication.

In this country we do not seem so concerned about the problem as they do in the United States. And we suffer for our apathy, as Cecil Chisholm reflects in *Communication in Industry*:

Evidence of poor communication is all around us. The average worker hasn't the haziest idea either of what his own job means or of what his company is up to, to let alone up against. Poor timekeeping, high rates of absenteeism, ceaseless changing of jobs, poor workmanship amongst youngsters especially, restlessness amongst juniors, the difficulty of the unions in getting their members to observe contracts; all these troubles, we are beginning to realize, are partly due to our failure to explain ourselves to our employees.

COMMUNICATION IN THE FIRM ON DIFFERENT LEVELS
The issues posed in the above are complex and wider than a mere matter of report writing. But this is the context in which reports are passing to and fro in firms. You do sometimes have to communicate with subordinates, you may be members of a large organization where face to face contacts are few, where the management superstructure is elaborate, and where you have to rely on the written word, the memorandum on the notice-board, the insert along with the pay slip.

In these circumstances how do you interpret? How do you determine the level of presentation and so of language? Some firms when communicating with all their employees make the mistake of issuing the identical message to shop floor, office staff, supervisors, and management, regardless of the fact that, say, a statement about a new pension scheme or holiday allowance will affect different categories in different ways. Very rarely will all of it be applicable. Striking the right level of presentation in such circumstances will be well-nigh impossible. Apart from the level of the language, supervisory and management grade might like to feel that their notification was in advance or issued separately from that to the shop floor.

Interpretation therefore is a matter of thinking less about your material and more about your readers. I suggested there were various types of readers, the administrators and executives, the specialist readers and general science readers, the supervisors, colleagues, and subordinates. Your understanding of their knowledge and needs, their possible emotional attitudes, and your relationship with them within the firm, must be the determining factors when it comes to writing out the first draft.

It is only with this appreciation of the total situation (the material, you, and the reader), that you can judge how much to explain and how much to take for granted ; how to strike a balance between condensation and tedious elaboration ; how to choose between technical and non-technical terms ; whether to use mathematics and if so how much ; what methods of illustration to employ ; what possible misconceptions to forestall.

USE OF ILLUSTRATIONS AS AID TO UNDERSTANDING
Diagrams and tables can be a great aid to understanding but only if the reader is equally familiar with their use as you are ; otherwise they can prove distracting and misleading. Professor Kapp puts it well :

Advertising agencies may be able to blind with science the gullible and non-scientifically-minded public, but a report writer should be very wary of using this technique. Quite often more mathematical information appears than is strictly necessary. This may be because the writer is unable to communicate in any other way, but sometimes it is a form of showing off, trying to produce in the mind of the reader the response, 'This may not be all relevant, and I am not sure I understand it all, but this chap certainly knows how to handle figures'. That may be a worthy motive for a student sitting an exam, and some examiners may fall for it. But the reaction of the average reader of a report is impatience.

The major difficulty of interpretation arises when the writer has to satisfy different levels of readers in the same report. The only guide here is to write at the level of the most important reader – the most important for you, not necessarily the most senior executive. It is far better to preserve consistency of style than to try to write at different levels, sometimes explaining what you feel to be obscure, sometimes not. This will be appreciated by your most rigorous reader. If possible, the introduction should be phrased in fairly general terms, so that the non-specialist can read it, and the technical detail can be reserved for the main argument of the report. Some people may appear on a circulation list for reasons of prestige, courtesy, or tradition (because no one ever deleted their name). These will not be your most critical readers.

However long a circulation list is – and some are a fair length – there will usually be three or four names who really matter. Write for these and take a chance on the occasional adverse comment of the personnel manager.

The last process in making the first draft of a report is presentation, which will be discussed in the next chapter.

3. The presentation and structure of reports

Presentation is to do with the actual lay-out and structure of the report and with such practical aspects as binding, spacing, and covering.

Matters of layout

These practical aspects should be self-evident but are frequently overlooked. Does your title appear at the top of a page of the text or on a separate title page? What difference does it make or does it not matter? What sort of paper do you use? What is the type in your typewriter? Do you have wide margins and double spacing? Is your report between stiff covers? How is it stapled? At the top or at the sides? Are the margins too narrow or the stapling too deep, so that the reader is missing the first few letters of each line or else has to bend back each page?

It is no good saying that you leave all these matters to your shorthand-typist. Many people do so with dire results. Nor is it any good saying that these questions are trivial, merely frills. Attractive lay-out and ease of handling pre-dispose the reader in your favour. No one likes reading closely printed text. No reader enjoys knocking his ash-tray over his trousers as he has to twist a report full circle in order to study a folded diagram. All you can do to make your reader's task more pleasurable in such small matters is well worth the small time spent on it.

Some of these considerations have a practical bearing. If your report is only to be seen once and then cast aside, the sort of cover it has is immaterial, but if your report is to receive a lot of handling, some protective covering should be afforded.

When making your decision about presentation, consider such matters as what will become of your report after it leaves your hands. What distribution is it to receive both within the company and outside? What arrangements have been made for coding and filing the report? Will it be possible in a year's time for someone to lay their hands on it within five minutes?

As for spacing and design and general lay-out, current practice is to be generous, almost prodigal with space. Some firms have their own policy on these matters and you are left with little freedom. The typists have all been trained in the firm's own secretarial school and know the accepted practice.

You obviously have very much control over a report if you write or type it yourself. Many people dictate their reports, and unless they give very precise instructions, the stenographer will interpret herself, sometimes not with outstanding success. Dictation can be a pernicious practice. It encourages verbosity and padding. A man writes differently from the way he speaks. Those reports which have a distinctive style are more often written than dictated. Some people prefer to dictate a rough draft then carefully shape and polish by hand. This method can provide good results.

Your main concern, though, will be the framework which you give to your report, namely its structure. This means in effect its headings.

Headings to a report

For a very short document headings would be superfluous and might even appear fussy and over-elaborate. For a document of any length they are essential. They are signposts enabling the reader to find his place quickly. They will be used in two ways, by people who do not wish to read the report but only pick out certain aspects from it, and by people who do read the whole report but wish to make quick reference to it some time later. The difference between these two classes of readers is that the second type of reader very often does not notice the headings at a first reading at a single sitting.

This is particularly true with articles in the technical Press. If you write an article for *Engineering* or for one of the professional magazines it will have a number of headings fairly regularly spaced. Not infrequently these will have been inserted by the editor. His caption may be a summary of what follows but often is a phrase taken from the text. The reason for this is to improve the lay-out of the page and to give a point of reference for subsequent usage. I think you will find that you are rarely aware of these headings at first reading. If you pause at all, it is but fleetingly.

The only criterion for a good structure to a report is, can the reader find his way about easily? The more casual reader of a report rarely begins at the beginning and reads the document carefully through to the end. He usually picks it up and scans it to see what he is in for. He looks for and interprets any signs that will help him to decide what the document is about, trying to decide how carefully and how much of it he will need to read, and what degree of urgency and concentration it demands.

Unrelieved typescript for page after page encourages the reader to postpone his reading of the report. Sub-headings, diagrams, tables act as a bait.

Possible structures of a report

There is no definite answer to the question 'What is the best form in which to design a report?' A Texas Oil Company boldly lays down in its Company instructions:

'A report should contain the following general subjects in the order mentioned: Introduction, Report Proper, Conclusions, Recommendations, Future Work, Appendix.

But not all reports would merit an Appendix. There may be no future work. The most appropriate structure will depend on the current practice of the firm, certainly, but will depend also on the experience of the writer and the purpose of the report. For example, the putting of conclusions and recommendations at the heading of a report before the main argument seems to some people to be putting the cart before the horse, illogical in the extreme. But to others who use the form – and many do – it is the obvious way of showing quickly the relevance and importance of the report.

You should adapt the structure of the report to fit the things you want to say. If you are asked to introduce a change of filing system or recommend a reorganization of the office, to start thinking in terms of the ideal layout, or of the Texas firm's directives, is to make your task unnecessarily difficult. For a report on this topic, the following structure might be the simplest: Present distribution of office furniture (which impedes efficient communication and logical flow of work); Proposed changes of office furniture; Advantages of change (and, possibly, disadvantages); Conclusions and Recommendations.

Depending on the length of the document you could decide whether you wanted an Introduction or whether a covering note would do.

Dear Bill,

Here is enclosed report on suggestions for re-planning Office Layout which you asked me to let you have by the end of the month.

Yours,

Fred

You might argue that the example I have given is an informal report and an even more informal note, but might this approach not be appropriate? There are as many informal reports written in firms as formal ones. If you are catering for the needs of a specific reader something over-elaborate might be quite unsuited.

Here are a number of standard structures prescribed by firms and textbook writers:

(A) Title page
Preface or foreword
Letter of transmittal
Table of contents
Abstract
Text of main body of report
Summary or statement of conclusions
Appendices
Index

(B) Title page
Summary
Introduction
Symbols
Description of Apparatus
Test Procedure
Analysis and discussion
Conclusions

      Appendices
      References
      Illustrative material
(C)   Title
      Table of contents
      Abstract
      Introduction
      Body
      Conclusions and recommendations
      Summary
      Acknowledgements
      References and biliography
      Tables and figures
      Appendices
      Index
(D)   Display cover and title page
      Table of contents
      Summary
      Introduction
      Conclusion
      Recommendations
         (Suggestions for future work)
      Discussion:
         1. Equipment
         2. Procedure
         3. Theory
         4. Results and their interpretation
      Appendices:
         1. Literature cited
         2. Detailed data
         3. Routine sample calculations which serve to sub-
            stantiate the result but which are unnecessary to its
            comprehension

Structures (B) and (D) look particularly appropriate to research reports, but what must stand out is the different ordering
of the same headings and the lack of an agreed system of nomenclature. (A field of growing concern is the international stan-
dardization of technical terms.)

Difference of terminology

Is a 'summary' the same as 'statement of conclusions', as structure (A) suggests, and a 'preface or foreword' the same as an introduction? How does a summary differ from an abstract, which is included in structures (A) and (C) but omitted from (B) and (D)? The Texas Oil Company which lays down precise instructions on the form reports should take, lays down equally emphatically that a 'summary is a separate part, not a component part, of the report and is for those not interested in the entire report. Such summary should make statements rather than discuss.'

Many people would say that this is precisely what an abstract is, that is, a detachable and separate part of the report which can be distributed independently of the main report. An example of an abstract is given on page 53. This confusion in terminology is not insuperable. As long as your reader is conversant with your terms it is not essential that all firms in the country should be.

Popular structure

The most frequent, most obvious, and probably most successful arrangement is the following: title page, summary, introduction, main text, conclusions and recommendations, acknowledgements, references, and appendices.

Some observations on the common terms used to structure a report would perhaps be helpful.

Title page

The title should preferably appear on a separate cover page. A title should be brief but not too brief. It should tell the reader what the report is about. Thus the report on page 55 has a reasonable title. 'Rip-Snorter Saws' by itself would have been inadequate. Similarly 'Report on Investigation into the Assembly of 1/4 B.S.F. Wire Thread Inserts' is preferable to merely '1/4 B.S.F. Wire Thread Inserts'.

Other information which it is customary to put on the cover page is the date of issue, the circulation list, the name of the author or authors, and the authority for circulation.

Table of contents

You have to use your common sense about the inclusion of a table of contents. If the report is only a matter of two or three pages, a table of contents would be unnecessary, almost pedantic. A document of some length would greatly benefit by having one.

An index would be even more of an extravagance for most reports ; something more appropriate to a text-book. The essential difference between the two is that a table of contents is chronological (and possibly categorical and summarizing), an index alphabetical.

Summary

The summary should be a survey of the ground covered in the report, brief but sufficient to indicate the area and depth of the study, and whether its objective was achieved.

It should not be regarded as part of the report but written *after* the report is complete. Its writing is a good test of a good report. If you can look back on your efforts, which may have been piecemeal, and state effectively what it is that you have discussed in the report, it augurs well.

A summary should be a real summary, and so more than ten per cent of the length of the report would generally be too long. Although brief and factual, the summary should not be in telegraphese. It is the reader's introduction to your report and you would want that to be favourable. The order of presentation in the summary ought to correspond with that in the actual report.

Summaries should contain no material not mentioned in the report itself. Summaries too often contain afterthoughts or statements to tidy things up, for which there is no evidence in the report proper.

Where the summary appears, whether at the beginning of a report or at the end, must depend on the practice of your firm. Some firms may want an abstract in addition to a summary, for additional distribution.

Specimen Abstract of Report

REPORT ON FAILURE OF RIP-SNORTER SAWS

EQUIPMENT IN USE:	RIP-SNORTER SAW Black and Decker N.75. Cost new £27 0s. 0d.
Motor details	220–50 volts, 50 cycles, 5.2 amps. Type A. Single phase. Series wound, 3,200 r.p.m. free speed at blade.
Blade details	No. 20361. Metal cutting saw. $7\frac{1}{4}''$ diameter. Fine tooth. 8 teeth/inch. Cost 26s. Cutting capacity $\frac{1}{16}''$ maximum thickness.

CONCLUSIONS:
1. Burning out of Rip-Snorter motors has been caused by:
   (i) Using blades beyond the designed cutting capacity.
       e.g. Universal use of No. 20361 blade.
   (ii) General mishandling by operators.
       e.g. Excessive cutting feed.
   (iii) Poor maintenance.
       e.g. Rewound armatures and inconsistent lubrications.
2. There has been frequently a shortage of these Rip-Snorter saws, because of their continued withdrawal for repair. The position will be improved by the introduction of blade number U. 1707, and by the use of paraffin as a lubricant.
3. As the applications of these saws are many and varied, discretion will have to be exercised by the operator in the choice of blade to be used.

RECOMMENDATIONS:

1. Stocks of 20 No. 20361, and 20 No. U.1707 blades must be built up in the Tool Stores to meet present requirements.
2. Worn undamaged blades must be returned to Black and Decker in batches of 6 for replacement at 4s. each. (New blades cost 26s. each.)
3. New armatures from Black and Decker must be used when repairing Rip-Snorter motors. (New armatures cost £4 12s. 6d. each, rewinds £4 0s. 0d. each.)

Approved by: A. J. HOBSON. Written by: H. R. LANE.
Date: 23 September 1958.

Note: Loan copies of full reports are held by Production Research Department, Hatfield and those marked*

CIRCULATION:

Hatfield
Mr T. Gilbertson
Mr J. A. MacKenzie
Mr S. R. Rudge
Mr P. G. Hodgkinson
Mr P. W. Grigsby
Mr E. Walker
Mr B. James
Mr P. J. Amesbury
*Mr W. E. Gatford (4)
Mr R. F. States

Mr C. R. Smith
Mr R. C. Grinter
Mr D. R. Knighton
Mr G. W. Weeks

Chester
Mr A. Turner
Mr R. Francis
*Mr W. Johnson
Mr S. Bradbury

Portsmouth
Mr S. Statham
*Mr E. W. Babey

Christchurch
Mr G. Davis
Mr A. R. Taylor
*Mr R. H. H. Kingdon
Mr J. Coulbert

Specimen Report

PRODUCTION RESEARCH DEPARTMENT

Report No. 59

REPORT ON FAILURE OF
RIP-SNORTER SAWS

Written by: H. R. LANE

Approved by: A. J. HOBSON

for Production Research Department

Date: 23 September 1958

CIRCULATION:

Mr W. E. Gatford, Hatfield.

Mr W. Johnson, Chester.

Mr R. H. H. Kingdon, Christchurch.

File (2)

REPORT ON FAILURE OF
RIP-SNORTER SAWS

OBJECT

To investigate the reasons why the motors of Rip-Snorter saws continually burn out.

EQUIPMENT IN USE	RIP-SNORTER SAW
Black and Decker	N.75. Cost new £27 0s. 0d.
Motor details	220–50 volts, 50 cycles, 5.2 amps. Type A. Single phase. Series wound. 3,200 r.p.m. free speed at blade.
Blade details	No. 20361. Metal cutting saw. 7¼″ diameter. Fine tooth. 8 teeth/inch. Cost each 26s. Cutting capacity ³⁄₁₆″ maximum thickness.

ALLOCATION				
1 Rip-Snorter	to		Redux Department.	Mr Brown
1	,,	,,	106 Wing.	Mr Millard
2	,,	,,	S.A.D.	Mr Lowe
1	,,	,,	Wood Mill.	Mr Smith

A 9″ Rip-Snorter saw exists in both the Redux Department and S.A.D.

DESCRIPTION OF EQUIPMENT

This is a portable circular saw. The blade is geared to the motor, which lies parallel to the rotation of the saw. Suitable handles and guides are incorporated in the design. The model is obsolescent.

APPLICATION

These saws are used for trimming reduxed wing skin panels, which are 1″ thick in places, general cutting of light alloy sheet and skin panels.

REASONS FOR FAILURE

The blades in use with these units are of a fine tooth form, being designed for cutting moderate gauge sheet. For heavier cutting a coarse tooth blade is required. In using the finer blades the motors are being overloaded, due to the fact that heavy pressure has to be applied to the saw to achieve a reasonable rate of metal removal, which is further hampered by swarf clogging the teeth. The high current taken by the armature when the motor is thus semi-stalled causes it to burn out. This type of failure is common to all saws that go to the Plant Department for repair.

On examining discarded blades, cracks were visible at the periphery, which is the result of bad handling, twisting, etc. This confirms maltreatment of the tool, which would in turn overload the motor.

A Rip-Snorter saw was examined and found to be in very poor condition. The base plate was twisted due to worn parts. Therefore jamming could take place between the blade and the work during use.

The gear boxes on the Rip-Snorter Saws have been known to run dry. This causes increased friction which overloads the motor.

Burnt-out armatures in most cases are rewound, and many of these rewinds have been found to be inferior to new armatures. As a result they have quickly failed when the saw has been put back into service.

TRIAL OF COARSER-TOOTHED BLADE

A U.1707 blade of the following details:

$7\frac{1}{4}''$ diameter, 4 teeth/inch, metal cutting blade, cost each 26s., cutting capacity $\frac{1}{4}''$ maximum thickness, was put on trial in the Redux Department. It proved to be an all-round improvement over a wide range of applications, including the cutting of 1″ thick light alloy plate.

CUTTING CAPACITIES OF BLADES

The cutting capacities of the blades quoted by the Manufacturers:
Blade No. 20361 $\frac{3}{16}''$ maximum thickness.
Blade No. U.1707 $\frac{1}{4}''$ maximum thickness.
relate to plain plates unhampered by the attachment of stringers, etc.

These cutting capacities can be exceeded, if the feed rate is reduced in proportion to the thickness being cut. Panels which have stringers attached offer a greater resistance to the saw. Therefore again feed rate must be reduced when cutting through these stringer sections.

CUTTING LUBRICANTS

In general, paraffin as a cutting lubricant can be introduced, but its use in the Redux Department must be limited to panels that have *undergone* Redux process. As a lubricant it will help to eliminate swarf clogging the teeth and sides of the blade.

REPLACEMENT SERVICE FOR WORN BLADES

Black and Decker operate a replacement service for blunt undamaged blades. These blades will be replaced by resharpened ones, at the following rates:

$7\frac{1}{4}''$ diameter at 4s. each
$9\frac{1}{4}''$ „ „ 5s. 6d. „

CONCLUSIONS
1. Burning out of Rip-Snorter Motors has been caused by:
     (i) Using blades beyond the designed cutting capacity.
          e.g. Universal use of No. 20361 blade.
    (ii) General mishandling by operators.
          e.g. Excessive cutting feed.
   (iii) Poor maintenance.
          e.g. Rewound armatures and inconsistent lubrication.
2. There has been frequently a shortage of these Rip-Snorter saws, because of their continued withdrawal for repair. The position will be improved by the introduction of blade No. U.1707, and by the use of paraffin as a lubricant.
3. As the applications of these saws are many and varied, discretion will have to be exercised by the operator in the choice of blade to be used.

RECOMMENDATIONS
1. Stocks of 20 No. 20361, and 20 No. U.1707 blades must be built up in the Tool Stores to meet present requirements.
2. Worn undamaged blades must be returned to Black and Decker in batches of 6 for replacement at 4s. each. (New blades cost 26s. each.)
3. New armatures from Black and Decker must be used when repairing Rip-Snorter motors. (New armatures cost £4 12s. 6d. each, rewinds £4 0s. 0d. each.)

The above report is referred to a number of times throughout the book. It has a number of stylistic faults. It is a good example of a report which conveys its information under appropriate headings. Though short, it is easy to find your way about it. The writer has allowed the nature of the report to dictate the structure. The report provides a good example of a detachable abstract.

Introduction

Many writers look back to their schooldays when they think of introductions. Introductions to the weekly essay were exercises in introducing your reader gradually into the subject. An apt quotation, a story, an analogy were considered to be praiseworthy efforts. A blunt 'This essay is about a visit to the dentist' would have been frowned upon.

In writing reports the best introduction is to be very old-fashioned and obvious and to state in your first sentence what the report is about. An example would be: 'This report discusses the use of ammonia to eliminate acid dewing in the stoker-fired boiler.' You can be even more positive. 'This report proposes the case for a change from coal-fired to oil-fired boilers.' Editors of technical journals never weary of reminding their writers to state the object of their article or paper in the opening paragraph or paragraphs so that the reader knows where he is going and so can assess and judge what he is reading.

The next step is to provide the necessary background information – such as the sequence of past events leading to the present problem – which your reader will need, and which he may not know or clearly remember. You may need to mention the authority and terms of reference. The final report may be different from the original terms of reference. The danger for many writers is to sketch in the historical background to the subject first, and at great length. You should avoid this.

The final stage is to inform the reader how you propose to develop the subject under discussion, and to provide an outline. At the same time you should define any technical terms or words you intend to use in a special sense.

The introduction more than any other section demands a wide knowledge of those for whom the report is primarily intended. The introduction is the place for the broad, general view rather than the particular and analytical.

How long it is will depend on what the reader already knows. If he is completely acquainted with the case up to the point where the argument starts it will be short. If the report is on a subject with which he is unfamiliar, he will have to be briefed with a good deal of historical background. It is unwise when introducing historical background to be controversial or contentious.

The most important part of the introduction is to let your reader know as soon as possible what you are driving at, to see your line of argument. In the following extract it is difficult for the reader to locate the true issue and face up to it. The picture is confused by qualifications that have been introduced which might have been more properly located in the body of the report.

Tubeless Tyres Ltd have broached the possibility of tanker movement of butadiene from the Refinery to Liverpool, resulting in a lower delivered cost to them. We have concluded that resulting savings are probably not sufficient to justify the investment.

Of the other purchasers of butadiene, the India Rubber Co. at Barrow can be included in the West Coast tanker movement, and Bouncing Balls Ltd at Old Trafford can be supplied by barge, the latter being a very attractive proposal. Despite the fact that a portion of the investment required at the Refinery to serve Liverpool and Barrow is common with that for Old Trafford, the return on the Liverpool investment (if the Old Trafford shipment were approved) still does not significantly increase.

What does the writer mean? Have you to go back and read it again? If so, why?

It is possible to reduce to one or two clear propositions a whole mass of evidence. Suppose a customer complains he is having trouble with the lubrication system of his mining machine. The notebook evidence might be assembled as follows. Substantiating the complaint:

1. An excessive amount of oil was being used.
2. The oil filter on machine 1 was clogged with fibrous material.
3. The filter on machine 3 was excessively dirty.

Investigation revealed:

1. Deposit analysis of oil from machine 1 shows contamination by cotton fibres.
2. The transmission on machine 1 had been cleaned with cotton waste.
3. The oil filter on machine 2 is clean.
4. The filter handle on machine 3 contains dust.
5. Deposits from machine 3 contain dust.
6. The machines are not dust-tight.
7. The hydraulic clutches leak.

The issue for elaboration which you would present to your reader in the introduction based on these findings would be:

1. Oil deterioration is not the cause of poor machine performance. The oil has been contaminated.

2. The customer's trouble is caused by bad maintenance.

The body of the report

The classification of your material in your earlier stages of preparing the report will have some bearing on the body of the report. In this section the issues outlined in the introduction are elaborated. If the report is of some length, it will be necessary to have sub-divisions with appropriate headings.

The main requirements are that the development should be logical, that the evidence marshalled should be relevant, and the reasoning clear to the readers. I suggested earlier that the logical order is not necessarily the chronological order, although sometimes it might be. The logical order, if you are describing a mechanism, would be from the general (the formation of the machine and the physical principles on which it is based) to the particular (the details of the construction).

There should be a proper balance as well as a proper sequence of ideas. One section should not be developed to the exclusion of another merely because you are more interested in it. Information that is outside, or appears to the reader to be outside, the purpose of the report should be excluded.

Conclusion

The conclusion can summarize the discussion in the main sections. (The body of the report may contain conclusions, as the marshalling of the evidence may lead to a conclusion which is stated.) Hilaire Belloc's recipe for a lecture might well be followed in report writing. First of all tell your audience what you are going to tell them. Tell it them. And then tell them what you have told them.

The conclusion can also summarize the findings and inferences. A result is not necessarily a conclusion, though some inference is likely to be drawn from it.

The conclusion should not contain any new idea not previously introduced into the report. It should consist of firm

unqualified statements. It should conclude. It should be obvious
to the reader that he has reached a natural finish to the report.

Recommendations

Although recommendations appear separately from conclusions in some reports and are bracketed together in others,
logically they should go together. In this sense they are part of
the conclusion. The natural process in an investigation report
is to obtain some results, from which you draw conclusions,
upon which you base recommendations.

It may be convenient to list them under a separate heading
or it may be more tactful to hint at them in the conclusions.
Recommendations can be the most difficult part of the report
to write. You have to consider very carefully your relationship
with the reader. The reader may consider that he is the person
who makes the recommendations. In that case it would be unwise – and might even be impertinent – for you to suggest what
action should be taken. You should ask yourself, 'Who does
make the recommendations round here?' If you are that
privileged person, and you wish the reader to take a certain
course of action, what should be the tone of your recommendations? Tentative? Conciliatory? Aggressive? You must decide,
if hostility towards the recommendations is likely, what is the
best approach to adopt.

The Appendix

The appendix or appendices, for there may be more than one,
proves a convenient way of presenting detailed information,
particularly of a descriptive nature, which if inserted in the
main body of the report would interrupt the smooth flow of the
narrative. An appendix should contain material not strictly related to the main argument of the report, but which nevertheless is of interest.

The sort of material that is relegated to the appendix is
experimental results, statistical data, tables and graphs, corres-

pondence, and worked examples. The merit of separating an appendix from the main body of the report is that the reader is under no obligation to read it. Some readers will find it irrelevant; other will not. It is also possible to record up-to-date work. It is better to put it in an appendix than completely re-write the report which has been conceived and written as a whole.

As in the case of the summary, the style of writing should not be different in the appendix from that used in the main body of the report. Page numbering should be observed. If there is more than one appendix, the other appendices should have an appropriate title and be designated as Appendix A, Appendix B, etc.

There are a number of other items which, while not an essential part of the *structure* of the report, are very often used, such as references, footnotes, acknowledgements, bibliography.

Acknowledgements

Most of us obtain the help of other people in the compilation of a book and sometimes in a report. If that has been on a generous scale it is only courteous to acknowledge it. This can be done in two places, at the beginning of the report in a preface or separately at the end. Some people may have helped substantially; others may only have allowed the right of quotation. Both sources should be mentioned, one formally, the other perhaps more personally.

References

If you use other people's work, particularly written or published work, it is customary to make some reference to it, not only so that credit may be given to the person whose work you are using (for in some measure this can be done in acknowledgements), but so that your reader may refer to the quoted work for confirmation and further study.

For this you must give precise details. These should include:

author's name and initials, title of the book, paper, journal, or report, publisher, date of publication, place of publication, edition, page numbers of citation (first and last), price.

The reference details will differ slightly in the case of book, journal or periodical, and report. Journals may be assembled in volumes, so a volume number will be necessary. Although publishers publish periodicals it is not customary to include their name. For obvious reasons publishers' names would not appear on reports. Some books go into more than one edition. You should state which edition you have used, since your information may be based on material from a book which has been revised in later editions.

Referencing is not something which should be lightly undertaken. Accuracy is essential. It can be extremely annoying to the reader to go to great trouble to obtain a journal or a book and find that the writer of the report when referring to it has mistakenly put Vol. XIV when he meant XVI.

Even unpublished material should be acknowledged if it is felt appropriate to do so, e.g.:

Kapp, R. O., 'The New Tower of Babel'. Paper read to the Presentation of Technical Information Group, at University College, London, 31 Jan. 1962.

Referencing is dealt with very fully in *Better Report Writing*.*

You will observe that I have just used an (*) for marking a reference. Practice varies. British Standard 1,629, 1950 gives the accepted methods for setting out references in full and covers all forms of reference.

Sometimes a variety of printers' marks are used; sometimes numbering is preferred. The numbering can be per chapter with the references at the end of the chapter or consecutively throughout the book with references at the end. The reference number is usually put at the end of the sentence. If a printer's mark is used, the reference will be found at the bottom of the page on which the reference is made.

---

*Waldo, W.H., *Better Report Writing*, Reinhold, New York, 1960, 2nd ed., pp. 41–54

Footnotes

Footnotes are more commonly to be found in books than reports. As their name implies, they are to be found at the foot of the page. You should use a different printer's mark for a footnote than is used for a reference. A footnote could be described as a very small appendix. Its insertion in the text would interrupt the smooth flow. You may feel you would like to make an aside or an elaboration. A footnote provides a convenient place to do it.

Footnotes can be very distracting. Unlike a reference, they have to be read separately rather than taken in with the sweep of the eye. The reader's attention is thus dragged down from the middle of the page to read an extract which may not prove all that relevant and illuminating. He then has to find the sentence which referred to the footnote and resume his train of thought. Many readers, myself among them, prefer to read the notes after they have read the book. Footnotes should be used sparingly and they should be brief. Those which take up half a page, as some of them do, are to be deplored. The matter, if that important, should have been incorporated into the body of the text or relegated to an appendix.

Bibliography

A bibliography can be a list of the books to which reference has been made or can be a list of additional selective reading. You should tell your reader whether your bibliography is a list of references or a guide to further reading. It is placed at the end of the book or report. Sometimes it is the practice to subdivide it into the different aspects of the subject, with short synoptic notes. A good example of this is to be found in *The Technical Writer*.*

*Godfrey, J. W., and Parr, G., *The Technical Writer*, Chapman and Hall, London, 1960, 2nd ed., pp. 318–33.

Numbering

It is common practice in many firms to emphasize the sub-divisions of a report by the use of numbers. Different orders of numbers can be used to indicate different degrees of sub-heading. In the Introduction a writer may inform his reader that his material is best considered under five headings. The numbering of the report will be from one to five or one to six, depending on whether the introduction is called Number One. Some writers, regardless of whether they announce in their introduction that their report is in so many sections, will automatically number all major headings.

If the writer is the type who feels impelled to number or categorize everything, his initial numbering in all probability will be Roman numerals (I, II, III, IV, V). His sub-divisions will be, first, Arabic numerals (1, 2, 3, 4, 5), then small letters (a, b, c, d, e), and finally, Roman numerals in lower case (i, ii, iii, iv, v).

Alternatively he may prefer the decimal system, starting with a 1 and numbering succeeding sub-paragraphs 1.1, 1.2, and so on. Sub-sub-paragraphs would have a second figure: 1.1.1, 1.1.2. . . .

Sometimes only sections will be numbered in this way; at other times each paragraph is so numbered. If this is the practice in your firm it would be wise to conform to it. If it is not, it is worth asking yourself whether numbering carried to this extent is not over-elaborate and, frankly, unnecessary.

You should ask yourself how much numbering helps the reader to find his way about the report. Is it any use? When a report has a very closely knit structure and classification, your reader can refer to a point by saying, 'I disagree with 2(c)'. Equally well he could say: 'I disagree with your third reason for failure given on page 9.'

In the desire for close numbering it is frequently forgotten that the prime method of referring to something is by the page. Very often this is enough. The reader will say (over the telephone, possibly), 'On page eighteen in the third paragraph under the section "Computer Explosion and Loading Procedure", I do not understand your line of argument.' You have to listen over the telephone much more carefully if your reader is saying 2.1.1 or 1.1.2.

The report on page 55 is a report which forsakes numbering almost completely (though 'Reasons for Failure' might with profit have been numbered), yet you would not say that it is a report in which it is difficult for the reader to 'find his way around'. Admittedly it is a short report, but those addicted to the decimal system would have profusely sprinkled even such a short report as this with 1.1.1, 1.1.2, etc. Quite a number of government publications, such as the report of the Royal Commission on the Press, number each paragraph, but this numbering is consecutive right through the report, ignoring the chapters and sub-divisions which the report may have. In all likelihood the report will finish up with a number such as 589.

Although numbering can help as a quick guide, it should not replace good headings. It is these that will enable the reader to pick up the gist of the report quickly. If he wishes to make a precise reference in a long report some weeks or months later, a page, a heading, and a paragraph reference may prove enough. But those who work for a firm which practises the decimal system should follow it.

Numbering of illustrations

The treatment of illustrations is dealt with in a separate chapter. Just as some reports lend themselves to appendices, others require illustrations. Their structure would be deficient and incomplete without them. The advantages and disadvantages of placing them opposite the text or at the end of the report will be discussed in Chapter 7. If you are writing a handbook and your illustrations appear throughout the chapters, you should be aware that the practice of numbering them consecutively is giving way to numbering them per chapter. The advantage of this is obvious. It allows new illustrations to be introduced in subsequent editions without necessitating the re-numbering of illustrations throughout the book.

*

Headings

The structure of a report as outlined in the previous pages
is only a framework, the basic shape on which a report is made.
It is not essential, nor one hopes, likely that the exact nomenclature would in all cases be observed. To have as a heading
'Main Body of the Report' is not very meaningful. Headings
should be as specific and illuminating as possible. The report
already referred to on page 55 has headings which reveal
their meaning straight away. The use of the heading 'Object',
is here preferable to 'Introduction'.

One report I received had four headings, 'Introduction',
'Comment', 'Additional Comment', and 'Conclusions'. Such
headings meant I had to read the whole report which, apart
from the inadequacy of the headings, was well written. Headings which gave some indication of what was to follow would
have saved me this trouble.

4. What is correctness?

You will at some time have been told that a particular sentence or phrase that you have written is incorrect. This is not the same as being badly written, which is a matter of style and, possibly, clarity. Style, we all know, is a matter of taste and, though there is a fair measure of agreement amongst critics as to what constitutes good style, there will never be complete agreement.

To be told that something is incorrect, however, implies an accepted standard of correctness, some authority to which to appeal. The engineer establishes his correctness by an appeal to the slide rule, the micrometer, the logarithmic tables, or laws which have been based on scientific observation and measurement.

Correct behaviour

The sources of authority in the matter of correct English are not so universally valid. Correct English is bound up with a whole code of correct behaviour. Lord Chesterfield in *Letters to his Son*, written in the eighteenth century, laid down in minute detail how his illegitimate son, Philip, should comport himself if he hoped for a career in the diplomatic service. Even such details as how to cough, how to laugh, how to enter a room, did not escape his attention. A glance round a second-hand bookshop will reveal a wealth of books, written in the early nineteenth century, on canons of taste, such subjects as 'How to propose to a young lady in marriage'. The purpose of these books, as of the Essays written a century previously by Addison, was to educate and consolidate the manners of the emergent middle class.

Great wealth, vast estates, or noble lineage could not differentiate them from the artisan. These were the prerogative of the landed aristocracy. Elegant manners could go some way to

mark the difference. The enormous expansion of the public
schools in the nineteenth century had one result which further
widened the gap, namely correct accent, which is referred to
colloquially as 'talking posh', and phonetically as 'received
pronunciation'. Professor Higgins, in Shaw's *Pygmalion*, realized he could never pass off Eliza Doolittle as a duchess so long
as she dropped her aitches. Before the war it was virtually impossible to obtain a broadcasting job with the B.B.C. such as
announcing or newsreading if you had any sort of dialect
accent. Accent, incidentally, has always had more social reper-
cussions in this country than in any other. As George Orwell
records, the Parisian proletariat would accept him, whereas the
London proletariat would not because of his Eton accent.

Such barriers are breaking down, and today on the Third
Programme you will, as often as not, hear professors discourse
on some abstruse subject in earthy, north country accents. And
Jim Dixon, in Kingsley Amis's *Lucky Jim*, manages to get by
in a provincial university with quite uncouth behaviour. Quite
a lot of contemporary drama is based upon revolt against
middle-class manners. Those weighty tomes on 'Etiquette', outlining in detail the correct behaviour to adopt when leaving
visiting cards or even how to set cutlery, are seldom thumbed.
Dress and manners are more casual nowadays.

Correct written English

It is worth pointing this out, as correct written English owes
something to general ideas on correctness as a whole. The
letters of Colonel Hutchinson, a highly educated man who
fought in the Civil War in the New Model Army, could contain the most idiosyncratic spelling. Nor was he the only one
to spell in an individual way. Yet today you may feel that there
are correct ways of spelling. Your conviction may be under-
mined by coming across, particularly in engineering and
scientific textbooks, such variations from the norm in this
country as 'center', 'program', 'gray'.

If someone today challenges the correctness of what you
write, your appeal is to such books as the *Oxford English Dic-*

tionary, Fowler's *Modern English Usage*, and one or two similar handbooks, even Sir Ernest Gowers's *The Complete Plain Words*. A dictionary however, when all is said and done, is only a guide, as a glance at one or two different dictionaries will quickly show you. You should remember that by the time a dictionary is published it is already a few years out of date. The same stricture applies even more to textbooks on grammar.

What the lexicographers and grammarians of these books record is *the accepted usage of the most educated people of the day*. Mark you, this cannot be said of a great many grammar books still in currency in some schools, in which the writers are obviously more knowledgeable about Latin grammar than English, for they spend pages on such matters as gender and case, which are largely irrelevant in English (though not in French or German). In Shakespeare's day there was little that we would recognize as grammar. The grammar schools that were founded during his time were to study Latin grammar. It was only years later that grammarians encrusted Shakespeare's magnificent English with a set of rules.

The important words of the first sentence of the preceding paragraph are 'usage' and 'educated'. You must realize that English language is dynamic rather than static. Usage is changing, imperceptibly in many cases, all the time. Fresh words are being coined. Words which were used only as nouns are being verbalized. Familiar words are being given an added meaning by a daring metaphorical use. And the people who set the fashion are writers, politicians, civil servants, the B.B.C., schoolmasters, university teachers. It is the currency of language of these people that the lexicographers and grammarians record. Sociologists may trace the speech patterns of the Cockney, but unless the Cockney's mannerisms are used by the top people they will not become established as *correct*. This is not to pass a value judgement on the dialect of the Cockney or the Liverpudlian. It could be argued in some ways that their speech pattern is richer, homelier, more immediate in its impact. It is merely that when we talk of correctness, for better or for worse, it is of the speech of *formally* educated people.

Spelling

Not all university dons are pedantic about correctness. Professor C. S. Lewis, the Cambridge don, wrote to *The Times Educational Supplement* about the proposed spelling reform in a way that will win sympathy from many an engineer.

Sir,

Nearly everything I have ever read about spelling reform assumes from the outset that it is necessary for us all to spell alike. Why?

We got on for centuries without an agreed common orthography. Most men of my age remember censoring letters of soldiers and know that even the wildest idiosyncrasies of spelling hardly ever made them unintelligible. Printing houses will always have, as they have now, their own rules, whether authors like them or not. Scholars, who know the ancestry of the words they use, will generally spell them accordingly. A few hard words will still have to be learned by everyone. But for the rest, who would be a penny worse if *though* and *tho, existence* and *existance, sieze* and *seize* were all equally tolerated. If our spelling were either genuinely phonetic or genuinely etymological, or if any reform that made it either the one or the other were worth the trouble, it would be another matter. As things are, surely Liberty is the simple and inexpensive 'Reform' we need? This would save children and teachers thousands of hours' work. It would also force those to whom applications for jobs are made to exercise their critical faculties on the logic and vocabulary of the candidate instead of tossing his letter aside with the words 'can't even spell'.

It is not only American spelling of English words that is in a state of transition. If you are to consider correctness only, which is the correct way to spell the following pairs of words: inflexion or inflection; disk or disc; inquiry or enquiry; install or instal; cipher or cypher?

If correctness rather than current usage is what is being sought, the first word given of the pair gives the more orthodox, preferred forms. You will protest, quite rightly, that you more often see *inflection* than *inflexion*, which goes to show that usage is a better guide than the book of rules.

And who knows where they are in the spelling of words ending in -ise? How do you spell such words as televise, supervise, advertise, magnetise, equalise, recognise, economise, criticise, oxidise, to name but a few. There is an increasing tendency for them to be spelt the way they are written above. In point of fact only the first three are spelt with -ise; the remainder should be spelt with -ize.

As a final illustration of the irrelevance of spelling to meaning, I am including a short essay written for me by a not very bright second year Ordinary National Certificate student. He was asked to write about the qualities needed by a Works Manager.

The first quality needed by a works manager is a througher understanding of his job. This understanding can only be gained by years of experience working with the men under him and apreciating their difficulties. The good works manager must have imagination and drive and be quite prepaired to submit to drastic changes in methods of production in order to increase output. The best way of maintaining good output from the man at the bench is to treat him as an individual and not just as another pair of hands.

The works manager has to keep the directors satisfied as well as the workers so he must be a fairly good negotiator particulary nowadays when many strikes are caused by very poor liason between the shop floor and the managers. Many of these unofficial strikes seem to be over very petty greviences but if they are investigated a little closer many turn out to be caused by complete misunderstanding on both sides because the liason between the men and the managers is almost non-existent. A works manager must therfore be able to speak well to both workers and directors and to do this he must be well aquainted with the situation of the men at the bench. A good sense of justice and fair play is very necessary in today's factory.

This is very much clearer than many reports one reads. It says directly without affectation what the boy observes, feels, and deduces. Many of the errors are of the kind that his professional superiors would make; the others do not impair the meaning of the passage. Most employers should be only too glad to have such a mature young student in the firm.

Capitalization

The uncertainty governing correctness in the matter of spelling applies also to such matters as punctuation, hyphenation, and capitalization. I have seen, for example, bunsen burner spelt 'Bunsen Burner' and 'Bunsen burner' as well as the way I have spelt it. He would be a foolish man who would spend much time arguing the merits of the correct form. The golden rule to adopt throughout is to be consistent. The chances are that your errors are less open to detection. Be bold and you stand a good chance of getting away with it. If you argue that Bunsen must be capitalized because it is the name of a person, you should on this principle capitalize Diesel, but more often you will find diesel engine spelt thus. A word derived from a proper noun tends to drop the initial capital when its use spreads. When in doubt do not capitalize. Capitalize, however, all important words in titles, division headings, sub-headings, and captions. You should not capitalize prepositions, conjunctions, and articles, though. Finally, capitalize figure, table, and volume numbers.

Punctuation

Study for a moment Figure 3. Most people would agree that Vickers Limited are a well-known, important firm. The extract given is from the annual report to the shareholders and so we might expect the firm to be on their best behaviour, at their most correct. A cursory glance will reveal that punctuation marks have been virtually dispensed with. Is the communication less clear as a result? The authors have obviously been more concerned with lay-out and felt that too much punctuation would have been fussy.

I will discuss in a later chapter the important function of punctuation as a guide to meaning. Here I am concerned with what might be described as conventional punctuation. How do you, for instance, write the date at the heading of your letters? Is it 1st January 1963 ; 1st. January, 1963. ; 1st, January, 1963. or one of the many other variations it is possible to construct?

Vickers Limited

DIRECTORS THE VISCOUNT KNOLLYS GCMG MBE DFC
Chairman of Board
MAJOR-GENERAL C. A. L. DUNPHIE CB
CBE DSO *Managing Director*
THE LORD BICESTER
A. O. BLUTH
SIR SAM H. BROWN
SIR GEORGE R. EDWARDS CBE BSC FRACS
E. O. FAULKNER MBE
A. H. HIRD ACGI BSC MIMECHE
G. H. HOULDEN CBE MINA
COLONEL A. T. MAXWELL TD
SIR THOMAS R. MERTON KBE FRS
SIR FREDERICK PICKWORTH
D. L. POLLOCK
E. J. WADDINGTON ACA

AUDITORS Peat, Marwick, Mitchell & Co.
Deloitte, Plender, Griffiths & Co.

BANKERS Barclays Bank Limited 54 Lombard Street,
London E C 3
Glyn, Mills & Co 67 Lombard Street, London
E C 3
Midland Bank Limited 5 Threadneedle Street,
London E C 2

REGISTRAR Glyn, Mills & Co 67 Lombard Street, London
E C 3

REGISTERED OFFICE Vickers House, Broadway, Westminster,
London S W 1

Figure 3

Quotation marks

There is certainly no general agreement on the use of single or double quotation marks, nor on whether the full stop precedes or follows the final quotation mark. If you are writing for a publishing house, it does not really matter which you use, as the established House Rules will be followed by the printer. Practice does vary amongst publishing houses. If the author wishes, it is possible for him to designate how he proposes to use single and double quotation marks. The authors of *The Technical Writer* use single quotation marks for ordinary words used in an unusual sense and double marks for jargon or unfamiliar technical terms. The examples they give are as follows:

The cell contains a "depolarizer", or compound which has an affinity for nascent hydrogen.

The longer the path traversed by the beam, the greater the 'splash' on arrival at the screen.

The best guide is to dispense with them wherever possible. This can be done in print, though not on the typewriter, by putting the quotation in italics or smaller type. Sir Ernest Gowers quotes with approval the comment from the *A.B.C. of English Usage*:

It is remarkable in an age peculiarly contemptuous of punctuation marks that we have not yet had the courage to abolish inverted commas.... After all, they are a modern invention. The Bible is plain enough without them; and so is the literature of the eighteenth century. Bernard Shaw scorns them. However, since they are with us, we must do our best with them, trying always to reduce them to a minimum.

Abbreviations

The practice which Vickers adopted in their report to the shareholders, of omitting punctuation marks which do not add to the meaning, is one that is obtaining increasing currency in abbreviations. The initials of a group of words which together make up a word which it is possible to pronounce have for some time lost the intermediate full stops, e.g. UNESCO, NALGO ; but even initial letters which do not make up a pronounceable word are tending to lose the intermediate full stops; ITV, TWA, and WEA.

Abbreviations of scientific units are commonly printed without the full stop, though there is no standard practice. Given below are two lists of abbreviations, one approved by the British Standards Institution and one used in America.

B.S.I.	AMERICAN USAGE	TERM
cal	cal.	calorie
c/s	cps	cycles per second
dB	db.	decibel
dia.	diam.	diameter
e.m.f.	emf	electromotive force
eV	e.v.	electron volt
f.p.s.	fps	feet per second
f.c.	ft.-c.	foot-candle
h.p.	hp	horse-power
h	hr.	hour
lm	l	lumen
p.f.	pf	power factor
r.m.s.	rms	root mean square
VA	va	volt ampere

Frequently it is better not to abbreviate. You should particularly spell out short words. Abbreviations for units of measurement should be used only when preceded by an exact number, e.g. 12 ft. If you write a few feet this should not be abbreviated. The abbreviation should be written in lower case unless the term abbreviated is a proper noun, Å (1) Ångström Unit. You should abbreviate titles only when they are followed by a proper name, Prof. S. L. Brown. If you have to repeat a phrase many times there is some justification for abbreviation – pounds per square inch, p.s.i.

Most people would caution against the use of an abbreviation which is the subject of discussion. Minor rules are: do not add an -s to form the plural of an abbreviation, do not punctuate an abbreviation unless it is identical with a word.

Hyphens

To talk of the correct use of hyphens is to invite difficulty. Nowhere is uncertainty greater. The authors of *The Technical Writer* give a list of preferred hyphenations recommended by the Institute of Physics. It includes such words as: boiling-point, cotton-wool, to-day, cathode-ray, horse-power, to-morrow, screw-thread, short-circuit, test-tube.

I have asked numerous report-writing classes of engineers and scientists to write down the correct usage of these words. Never once have I recorded more than sixty per cent correctness from anyone by the Institute of Physics' standards, and never uniformity.

The dictionary would not agree with some of the Institute of Physics's preferences. The dictionary I use does not hyphenate boiling point, ground level, today, tomorrow, screw thread. Some American books spell test tube and shortcircuit without the hyphen; some spell wave length as two words instead of one; some combine horsepower, on the other hand, as one word.

The hyphen is used to combine two words to make a single compound but it is difficult to trace any pattern. Why should, for example, house-agent be with a hyphen, house master without, and housemaid one word? The tendency generally is for words to be first used separately, then combined by a hyphen, and finally made into a complete word.

The use of the hyphen is only mentioned now to indicate the state of flux which obtains in current practice. As Fowler says: 'The chaos prevailing among writers and printers or both regarding the use of hyphens is discreditable to English education.' The person who insists on dictionary correctness in this matter is a pedant. A few rules people will agree on, e.g. it is incorrect to hyphenate adverb-adjective combinations such as 'popularly known', 'recently acquired', but otherwise there is little uniformity.

Numbers

There is as great a lack of unanimity about numbering as there is about the use of hyphens. Should you write numbers as numerals or as words? Your best guide is clarity and appearance on the paper. Does your text look like an enormous equation or, in contrast, are some important figures unnecessarily disguised in a forest of prose?

Here is a list of fairly obvious rules:

Do not use numerals at the beginning of a sentence.

Do not use two numbers in succession (write 20 four-cylinder engines, or thirty 12-in. bolts).

Do not use numerals for round number estimates or ordinals (write, 'approximately two hundred parts were delivered', 'this was the third report on shortages').

Use numerals for all page numbers, dates, figures, diagrams, addresses.

For numbers below ten, words tend to be preferred to numerals.

Grammar

Some of the rules of grammar are no longer defended with the same rigour. The absurd one which maintained that sentences should not end with prepositions received its death blow at the hands of Sir Winston Churchill.

Another equally antiquated rule is the one that says sentences should not begin with the word 'and'. Although the persistent use of this word at the beginning of a sentence would be bad, because monotonous, it is quite legitimate on occasion to use it.

The following four sentences, however, would be regarded by most people as incorrect. Are they incomprehensible though?

(a) The number of jigs provided are adequate.

(b) The reason why he left us was because he felt afraid.

(c) Your theory is quite different to mine.

(d) We could do nothing to prevent him leaving.

(a) is incorrect because you have a plural verb, 'are', for a singular subject 'number'. (b) should read either 'the reason why he left us was that he felt afraid' or 'the reason he left us was because he felt afraid' – the use of 'why' and 'because' together is redundant. (c) is incorrect because it is customary to say 'different from' (but indifferent to). And in (d) the writer has failed to realize that 'leaving' is a gerund or a verbal noun, and therefore the pronominal adjective 'his' should be used and not the pronoun 'him'.

Unrelated participle

The misuse of the participle is perhaps the most common grammatical error to be found in technical writing. Its misuse arises principally when the writer tries to write in the third person and the passive voice. How many times does one read such sentences as:

'Using a different alloy, it was found that . . .'

'Running the machine for eight hours, the heat proved too much.'

'Having satisfied himself that these are correct, it is equally important that the airframe is correctly dimensional to receive the components.'

So frequent is this usage that I am tempted to say that it has almost become correct. Or, put another way, it has become accepted by readers of reports. The incorrectness, of course, lies in the fact that the present participle must have a subject to agree with and that subject must be the agent performing the action. If personal responsibility is not allowed to obtrude into a report by the discreet use of the first person, 'I' or 'We', the result will be the frequent use of the unrelated participle. I will discuss in a later chapter where I think the first person might reasonably be used. Similarly misused is the unrelated infinitive.

'To determine the correct temperature, readings must be taken.'

Is this correct or not? Do you understand it?

What you can get away with

If correctness is your only guide when you judge writing, what would you make, for example, of the extract from an article in the *Spectator,* a literary journal which might be said to set the fashion in accepted usage?

It has been suggested by Mr A. J. P. Taylor, in one of his *Sunday Express* exercises in how to be readable but wrong about world affairs, that Mr Nehru's brave words are nothing more than an astute move to enlist Mr Eisenhower's sympathies. That he is working on the appreciation that unquestioning support by the President for India against China would make up for the political difficulties at home that arise out of his amiability towards Mr Khruschev. All designed to extract economic aid more easily from the Americans. . . .

If Mr Eisenhower's visit to India helps him to present Indian poverty so much more dramatically to the American public that Congress will think twice before cutting down his request, and perhaps even reverse its decision that loans for development must be spent in the United States – an uneconomic piece of big-business selfishness – it will be a trip worth while.

Which is more than can be said for what is to be (except for a luncheon engagement in Morocco) the last visit of all, to Madrid, where the President is to drive through the streets with Generalissimo Franco, escorted by cavalry – no doubt by representatives of those same Moorish regiments that the Generalissimo brought to Spain to help in putting down its lawful government. . . .

I am not concerned with the political views expressed, but with what the writer has been able to get away with. A sentence is often defined as a group of words that makes complete sense. The second sentence in the first paragraph certainly does not make complete sense taken by itself, but only gets its meaning from the context. It is in fact a subordinate clause parallel to 'that Mr Nehru's brave words . . .' and might more appropriately have been linked by 'and' or 'or' – even a comma after 'sympathies'. It is not quite clear what 'all' in the third sentence refers to – 'Mr Nehru's brave words' or his words plus his actions in 'working on the appreciation . . .'.

The first word of the third paragraph should really be the

demonstrative pronoun 'this' rather than the relative pronoun 'which'. Who ever heard of a *relative* pronoun beginning a sentence, let alone a paragraph?

Is this bad writing? The test of good or bad writing must surely be that impact, of clarity, of ease of reading, not of strict adherence to conventional rules. Apart from the uncertainty of what 'all' refers to in the last sentence of the first paragraph, the whole extract is readable, clear, and states its case with some emphasis. The same criteria must surely apply to technical writing.

Changes in usage

I am not trying to make a case that all rules should be flouted, but to show that English in all its forms is fluid. The changes that are taking place are not so dramatic and obvious as the ones that you can see by comparing Old English, Chaucer, and Shakespeare, with language today. But an Old English word like 'hlaefdiġe' or 'wifman' only gradually became 'lady' and 'woman', as various sounds were dropped and altered. In those days pronunciation and spelling approximated more. Printing tended to standardize spelling, so that now we preserve the same form of spelling for words such as 'love' and 'prove' which we pronounce differently and yet we know that in Shakespeare's time they were pronounced identically.

> If this be error and upon me proved,
> I never writ nor no man ever loved.

And as late as the eighteenth century 'tea' was pronounced like 'day'. In educated speech today words such as 'psychiatrist' and 'calibre' are pronounced with the first syllable both stressed and unstressed.

The most important changes are usually extensions of the meaning of a word. Idiot, imbecile, feeble-minded were words used in mental hospitals to denote different types of mental deficiency. As these words have long been used as terms of insult and abuse, the Mental Health Act of 1959 decided as a result of the climate of opinion, which was sympathetic to all types of mental disorder, that they should be replaced by such neutral, inoffensive terms as severe mental abnormality, psychopathic disorder, and similar medical-sounding words. It only requires a number of people to use the word 'psychopathic' frequently and in a tone of scorn when they wish to convey their feelings of impatience about someone they momentarily dislike, and the word 'psychopathic' will become just another synonym for 'stupid'.

Mr Macmillan referred one day in 1963 to a question asked of him at the Manchester Cotton Exchange as 'the 64,000 dollar question'. This has become a phrase, one might almost say a cliché, for the most difficult question a person may be asked. Its origin was a question on a television show.

There are any number of words which we use metaphorically. A word such as 'square', used by teenagers to refer to their stricter elders, first appears in 'inverted commas'; its use may become so common that gradually the inverted commas disappear. New words, such as 'bikini' describing a brief swimming costume, no longer have inverted commas and are to be found in modern dictionaries.

What meaning do you give to the following two words: Happy – lucky? appropriate? felicitous? cheerful? Gay – light-hearted? carefree? immoral? showily dressed? dissipated?

All meanings are permissible. In the eighteenth century, 'gay' most frequently meant 'immoral'. Today 'light-hearted' is probably the most popular use, as 'cheerful' is the contemporary meaning for 'happy'.

The lesson of all this, therefore, is to respond to language as a living thing, to be sensitive to its changes and nuances and to preserve a freshness about your own writing.

The cardinal sin in all writing, but particularly technical writing, is vagueness. What contributes to this is not always neglect of 'the rules' but other factors which will be considered in a later chapter.

I would be wrong to finish this chapter if I were to suggest that you can flout the established conventions and make up your rules as you go along. A few novelists, such as James Joyce, can enjoy this luxury and get away with it. For most of us we have to observe certain minimum standards of correctness. You should be guided, therefore, by the dictionary and by the practice in the firm or of the professional body to which you belong.

You should bear in mind too that the ultimate judge of acceptability of your report is your reader. If minor errors of correctness distress him (or them), it is wise to reduce these errors to the minimum, even though their correction does not add materially to the comprehensibility or readability of your report.

*For a very full treatment of this subject of correct English there is no better book than Randolph Quirk's *The Use of English*, Longmans, London, 1962.

5. Style

Where reports break down, and where report writers experience most difficulty, is not in the structuring of the report or the uncertainty of whether to capitalize or not, but in the actual writing – in the style – of the report.

I suggested in the Introduction that it was mistaken to suppose that style was a kind of lacquer or garnish added afterwards to the plain statement of your first draft, in order to make the report attractive to your reader. It is something much more fundamental than that. It is very much to do with your relationship with your reader; but also with your sensitivity to the use of words and to the rhythms and structure of the English language.

I will discuss in the next chapter some examples of the personal style and of the diplomatic and phoney style, and the problems associated with them.

Whatever style you ultimately adopt, though, will find expression in words, in sentences, and in paragraphs. It will boil down to such matters as word order, punctuation, and whether you prefer the active to the passive voice. It is these and other matters which I would like to consider now.

Sentence structure

WORD ORDER It is not a matter of indifference whether you write:

'A new product must not only be created, but experimentally developed, to be successful.'

or

'To be successful, a new product must not only be created but experimentally developed.'

One gives a different emphasis from the other in the same way that 'Sir Alexander Fleming discovered penicillin' and

'Penicillin was discovered by Sir Alexander Fleming' are different.

The normal word order of a sentence is subject first, immediately followed by verb and then object or complement. This is the usual way of constructing a sentence, but it is often the least emphatic and therefore least effective way. Expression may be heightened by using an inverted word order.

'Kerosine is the safest fuel for use in modern jet engines,' is normal.

'The safest fuel for use in modern jet engines is kerosine,' is inverted word order. You must judge for yourself which method suits your purpose best. Which of the following makes its point best?

'Overtime went up as a result of the recent strike.'

'As a result of the recent strike, up went overtime.'

Simple sentences like those above do not present great problems. Difficulty arises when you introduce a number of qualifying statements which tend to extend the separation between subject and verb, or make the connexion between a relative pronoun and its antecedent more remote.

The following two sentences from an engineer's report are structurally awkward:

(a) 'Components are soaked here after washing in cold water to facilitate drying when removed.'

(b) 'The delay seems to be caused by the reluctance of the Machine shop to collect the castings as the necessary paper work has been completed.'

It is difficult to grasp their meaning at first reading. Surely a better way of writing them is:

(a) 'After being washed in cold water, components are soaked here. This facilitates drying when they are removed.'

(b) 'As the necessary paper work has been completed, the delay seems to be caused by the reluctance of the Machine Shop to collect these castings.'

Here are some more extracts that are poorly structured. They are taken from actual reports:

(a) 'Inspection of the stepped bolts that pass through the top boom of the centre section spar into the fuselage frame adjacent to the roof end was attempted but was found to be unsatisfactory and inconclusive.'

The trouble with this is that the subject 'inspection' is separated too far from its verb 'was attempted'. Two views can be taken about the re-writing of this: one, that the reader is familiar with the position of the stepped bolts and the description is unnecessary, or, two, he wishes to know where they are. In that case a better version would be:

(a) 'Inspection of the stepped bolts was attempted but was unsatisfactory and inconclusive. These stepped bolts pass through the top boom of the centre section spar into the fuselage frame near to the roof end.'

I am not sure that 'unsatisfactory' is a sufficiently precise word, nor is 'attempt' a happy choice. Attempt suggests that the author tried to do something but was prevented. What was really in the author's mind was that a superficial inspection had been carried out. It would have been more accurate to have said so. The expression 'found to be' does not add appreciably to 'was'.

(b) 'When drawings for tools to be subcontracted are received they should be studied by the Outside Representative so that he may decide which firms that he visits have the necessary equipment and capacity to undertake the work under consideration. In making the final decision consideration should be given as to whether the material required to manufacture the tools is held by the sub-contractor and has he successfully made this type of tool before.'

The first sentence is not altogether satisfactory but it will do. It could be made into two sentences, and the words 'under consideration' are superfluous. It is the second sentence of the original which falls to pieces. The first sentence uses the 'Outside Representative' and 'he', so why not continue in that vein:

'In making the final decision he should find out whether the sub-contractor holds the material required to manufacture the tools and whether he has successfully made this type of tool before.'

'Consideration should be given as to whether' is a most clumsy expression: the active voice 'holds' is preferable to the passive 'is held by'; and the parallel construction in the original is distorted by 'has he' rather than 'he has'.

(c) 'This examination was made in answer to a request by the Structural Engineer to determine whether a crack known to be in this area under a repair capping had extended.'

Do you find that easy to comprehend at first reading or is it a mouthful? The context of the sentence was as a part of an X-Ray report, a type of report which follows a certain pattern. The message might even have been conveyed in note or telegraphic form. In certain circumstances notes can be valid as a form of accurate communication. I give the whole report, leaving out the stereotyped headings such as Aircraft Type, Owner, Date of Manufacture:

Radiographic Examination of repair on stringer 26 between Ribs 14 and 15 on the starboard bottom mainplane skin.

This examination was made in answer to a request by the Structural Engineer to determine whether a crack known to be in this area under a repair capping had extended.

It was found that the crack had not extended beyond the stop hole.

X-Ray Section
Service Dept

What is the main sentence? 'This examination was to determine if a crack had extended' or 'This examination was made at the request of the Structural Engineer'. In the first example the subject has been separated by a detailed account of the location of the crack. Better to write:

'This examination was to find out if a crack, known to be in this area under a repair capping, had extended. The examination was requested by the Structural Engineer.'

One wonders whether 'in this area' is necessary as the area is defined in the title. It would be possible for reports of this type to be headed in a different way, under some headings such as: 'Examination called for', 'Authority for Examination', 'Result of Examination'.

(d) 'A test pressure ratio switch was set to a pressure ratio of 0·69 ... At this setting it was found possible just to stall the aircraft on to the ground without the device operating. This was thought to be due to a ground effect possibly resulting from flap blockage causing a loss of lift when very close to the ground.'

It is the last sentence which destroys the meaning. Too many consecutive and related actions have been run into one sentence, though it is not a long sentence.

It is also not absolutely clear in the last sentence of the extract what 'this' refers to. The failures of the device to operate or being able to just stall the engine? The misuse of pronouns is a very common fault in technical reports. Rather than use such expressions as 'it' and 'this' to refer to something mentioned a long time back, it is safer, at the risk of monotony, to repeat the subject.

The last sentence would read more clearly if phrased thus:

(d) 'The failure of the device to operate was thought to be due to a ground effect possibly produced by flap blockage. The flap blockage would cause a loss of lift when the aircraft is very close to the ground.'

(e) 'A series of stalls was carried out with engines idling in each configuration, taking continuous auto-observer records throughout.'

'Series' can be both plural and singular. The main fault is the unrelated participle 'taking'. Sometimes its usage can be forgiven, but not here. One feels that the writer was careless or lazy. Better to have written:

(e) 'A series of stalls was (were) carried out. Continuous auto-observer records were taken throughout.'

(f) 'In conclusion the Method Engineers responsibilities continue until the aircraft leaves the production line, which includes the incorporation of all modifications.'

The omission of the apostrophe from 'engineer' is not all that important, a matter of correctness. The 'which' surely refers to the engineer's responsibilities. If the writer wished it to refer to the production line, he would have to make a new sentence, such as 'The incorporation of all modifications takes place whilst the aircraft is on the production line' and so dispense with 'which' altogether.

An improved version might read:

(f) 'Finally, the Method Engineer is responsible for the

incorporation of all modifications to the aircraft till it leaves the production line.'

In making alterations to most of the above extracts I have tried to adhere as closely as possible to the style and structure of the author, making the marginal correction. If you take the following extract and compare it with an amended or alternative version, the style, the expression, the choice of words is different. One is not saying quite the same thing as the other.

1. 'In order to eliminate the effects of changes in the value of money over the period, the first costs of the older aircraft and their engines have been raised to conform approximately with present day prices of comparable equipment per lb. of airframe weight or per horse power.'

2. 'During this period the value of money has changed. The first costs of the older aircraft and engines have therefore been increased to eliminate the discrepancy between pre-war and modern prices.'

The title of the report was 'Progress in the Civil Airliner; 1935–60'. Hence the term 'pre-war' in the amended version. One feels that, though the second version has left out the term 'comparable equipment per lb. of airframe weight or per horse power', this is just a matter of detail. The impression of the second version is that this is lucid prose, not fine style, and that the author is just as likely to be technically accurate when it matters. More important, although the selection is probably too short to make such a comparison, a long report in the style of the second would be easier and pleasanter to read.

SENTENCE LENGTH One of the factors which affects the readability of a report is the number of words used in the sentences.

I have considered some of the ways in which word order can affect the meaning of a sentence, and suggested that the rearrangement of simple sentences is a relatively easy matter:

e.g. 'Overtime went up as a result of the recent strike.'

'As a result of the recent strike, up went overtime.'

Some of the examples I used as illustrations of faulty word order, however, required more re-writing than a mere rearrangement of the words in the sentence. The sentences themselves were too long and had to be broken up into smaller ones in order to reveal their meaning more readily.

Communication breaks down when you try to cram into your sentences too many subordinate ideas. It is easy to see how this happens. It is often difficult to make categorical statements, for no sooner have you made them than you feel you must qualify them in some way. 'No fracturing of the metal will occur during heating', must be qualified with 'provided the temperature remains constant'. The temptation is to go on and explain why this is so, and add some further subordinate clause, 'since big fluctuations in temperature. . .'. Nor is this the end. It is quite possible to build up more connexions which may be both logical and accurate in describing the experiment under discussion, but which are not easy for the reader to take in.

I suggest the comprehension of the following passage is about fifty per cent at first reading:

The preceding chapter has attempted to develop a fundamental conception of exposition as a kind of writing concerned with the orderly communication of ideas in accordance with a preconceived purpose and plan and of a report as a specialized form of exposition in which the use it is to serve determines almost everything – plan, scale, and method of treatment, and especially the amount of interpretative comment required.

It is the sort of paragraph which you have to read again in order to understand what the author is driving at. The main reason for this is that the sentence is too long. It also contains too many abstract words. Such words as 'preceding', 'fundamental', 'conception', 'exposition', 'communication', 'in accordance with', 'preconceived', 'specialized', and 'interpretative', do not leave a very clear image on the mind.

The passage could with profit have been divided into at least two sentences, with a full stop after 'plan' and the subject of the sentence repeated with some such wording as, 'It [the preceding chapter] has also considered the report as a specialized form of exposition. . . .'

It may have been a small point but an opening such as 'The chapter you have just read' rather than 'The preceding chapter' helps to personalize and prepare the reader for 'the fundamental conception of exposition', which is a phrase he will have to think about for a minute. Obviously you cannot avoid using abstract words – nor is it desirable that you should – but the over-use of them inevitably slows up the reader. If a book is full of such paragraphs as the above most readers will find the going heavy and lose interest.

READABILITY Some very interesting research on readability has been done in the United States. The Americans, with typical thoroughness and with their passion for organization, have established Readability Clinics. As far back as 1935 there was research into readability when William S. Gray and Bernice Leary brought up to date the work so far in their book, *What Makes a Book Readable*. The work on readability has been statistical. It has dealt with factors that can be measured, such features as how frequently particular words appear, what is the number of words per sentence, what is the number of polysyllables per hundred words, what are the levels of readership of particular magazines? Counsellors on readability do not try to measure the beauty of a particular piece of writing or whether the word order is the best possible, for the good reason that such things cannot be measured.

One of the most successful investigators into readability was Rudolph Flesch, whose books *The Art of Plain Talk* and *The Art of Readable Writing* had the same sort of impact in the States as Sir Ernest Gowers's *Plain Words* had in this country. His readability formulas were extensively used by government departments, where some publications were completely revised in the light of them. Flesch was called in as a consultant to make some of the forms and regulations intelligible to the ordinary reader. The U.S. Department of Agriculture even went so far as to start a Readability Laboratory to test all their own publications. The influence of Flesch and others has been more pervasive in the States (where they take up such studies with enthusiasm) than has that of Sir Ernest Gowers in this country.

Various readability yardsticks have been devised. The most recent and possibly the best is that worked out by Robert Gunning in his book, *The Technique of Clear Writing*. The devisers of such yardsticks are at pains to point out that these yardsticks are not guides to good writing, but means by which you can check up afterwards how well you have written. Objective, impersonal testing usually confirms the subjective impression that the extract is turgid and uninteresting. The testing is not meant to replace personal judgement.

Gunning's adjective to describe writing which makes heavy reading is 'foggy'. Hence his term, the Fog Index, to measure your readability. This is arrived at by the following three steps:

1. Jot down the number of words in successive sentences. If the piece is long, you may wish to take several samples of 100 words, spaced evenly through it. If you do, stop the sentence count with the sentence which ends nearest the 100 word total. Divide the total number of words in the passage by the number of sentences. This gives the average sentence length of the passage.

2. Count the number of words of three syllables or more per 100 words. Don't count the words (a) that are capitalized, (b) that are combinations of short easy words like 'book-keeper and butterfly', (c) that are verb forms made three syllables by adding -ed or -es (like 'created' or 'trespasses'). This gives you the percentage of hard words in the passage.

3. To get the Fog Index, total the two factors just counted and multiply by ·4.

The Fog Index is graded from six to seventeen and approximates closely to the American school-grade levels of reading difficulty. The latter have been calculated on the basis of questions asked of the student on reading selected passages to determine how well he has understood the passage.

Gunning regards a Fog Index of twelve as the danger point. If you rise above this, your writing is becoming difficult to read. A fairly exhaustive analysis of authors, past and present, classic and popular, shows them to have had Fog Indexes of under twelve. Gunning is able to report with satisfaction that a highbrow magazine which had a reading difficulty well above the danger line of twelve went out of business within a year, whereas magazines such as *Time* and *Look* which have a reading difficulty (or Fog Index) of ten and eight respectively remain in business with mammoth circulations.

Journals and authors who must please to live have to come down to the level of their readers. Writers of reports are fortunately not in the extreme situation of having to satisfy millions of readers to remain in business. If they were they would probably starve. People read reports because they have to rather than because they positively want to for the sheer pleasure of it. Although this eases the pressure, there is the incentive that people will read you with more eagerness if you pay some regard to their reading abilities. There is the comfort, too, that the Darwins, the Einsteins, and the Eddingtons have not only been great scientists, but scientists whose writing we have read with pleasure.

The background to the Fog Index and the statistical research behind the readability yardsticks need not concern you. The factors from which they are derived do repay study. The guide to good writing is not the yardstick but the importance of limiting length and not over-using what Gunning calls 'hard' words. He himself lays down ten principles of good writing, one of which is 'Keep your Sentences Short'.

Reports whose sentences *average* more than twenty words per sentence Gunning regards as difficult reading. All this analysis may appear very rigid. Certainly the eminent authors whom Gunning invokes to support his analysis did not solemnly sit down and plan to average twenty words per sentence. They planned to write as well as they could ; the result was that they produced this average. Any attempt to compose the passage in accordance with the formula is most likely to be unsuccessful.

KEEP SENTENCES SHORT On the whole, though, it seems clear that shorter sentences do make for greater comprehensibility. Some selected passages will illustrate this. Take the following extract:

From the formulae already derived, it is evident that the two-element type of low-pass filter sections can be combined with the mid-series equivalent or mid-shunt equivalent m-derived types of sections [in such a way] that the resulting structure will have a total image transfer constant which is the sum of the individual image transfer constants of the component filter sections, and image impedances which will be determined solely by the characteristics of the terminal sections of the composite filter.

Even a person fully conversant with the jargon would halt over that. Does the following version lose anything and is it not considerably clearer?

From the formulae already derived, it is evident that two-element low-pass filter sections can be so combined with mid-series or mid-shunt equivalent m-derived sections that the resulting structure will have the following features: its image transfer constant will be the sum of its component sections, and its image impedances will be determined solely by the characteristics of its terminal sections.

In the first version you feel that you are beginning to lose your grasp of the meaning by the time you have come to 'component filter sections', and you are wondering how much longer you can hold out before you can rest for a moment at the full stop, and so digest what you have read.

The colon in the amended version, which in this case is almost equivalent to a full stop, provides the necessary pause. You will also observe that the extract has been structured in a slightly different way. The second version with its 'will have the following features' gathers up what has been said and prepares you for what is to follow. About the second half of the sentence there is both a balance and a rhythm, 'its image transfer constant will be ... and its image impedances will be ...'. 'Combined in such a way' becomes more neatly 'so combined'. There are other small improvements which you can detect for yourself. Even with these improvements the passage still presents some difficulty.

WRITING WITH A HIGH FOG INDEX The following passages, taken from a variety of reports, are typical examples of technical report writing. Most of them betray the fondness for lengthy sentences. By the addition of one or two stops and some rearrangement they would yield their meaning sooner and with less effort for the reader. They are printed exactly as they appeared.

(a) 'When ordering single details, all reference to any scheme, other than the one being ordered for, will be blanked out or scored through on sheet 3 of layout, thus leaving only the information and quantity required for the particular scheme being ordered for the layout.'

(b) 'In all cases of chobert rivets not securing stiffener attachment angles, the attachment angle has been bent away from the adjacent structure thus making it impossible for the chobert rivets to engage in the holes in the attachment angles.'

(c) 'The tail loads produced by any oscillation in the elevator channel between the limit switches up to five seconds from the start of the oscillation are small compared with the design tail load.'

(d) 'The case has also been considered of the pilot winding on the trim until the fine limit switch operates flap up or to full extent and then disengaging the autopilot with flaps down (the coarse limit switch range is greater than the trim range).'

(e) 'On special jobs where the normal cutting oils will not suffice due to extremely heavy cutting pressure, i.e. breaching of high tensile steel, to produce the finish required this oil is used although it is extremely expensive.'

(f) 'There is, however, some reduction of the warning margin which is thought to be caused by loss of lift due to flap blockage if an attempt is made to prolong the float on landing when very close to the ground, but the margin is still considered to be adequate.'

(g) 'The main cause of failure is considered to be the exertion of unequal loads around the periphery of the glass during assembly of the masking frame, due to the following reasons.'

(h) 'From what has been seen during this investigation, it is most strongly recommended that the practice of scraping the casting, be immediately prohibited; other means being sought for making allowance for the known casting discrepancies, e.g. in 4.3 below.'

(i) 'At the start speed and time of contact were fixed while pilot material, lubricant, and load were varied, once pilot material and lubricant had been determined load and speed were varied and finally time of contact.'

(j) 'A smooth surface finish produced by either a rubber

abrasive wheel or 400 grade emery paper however carefully cleaned tended to promote scoring.'

(k) 'The report indicated by the ticks will be issued from the new Library, any not in it being immediately transferred from the old Library and all of these together with a copy of the list will be taken back to the member of the staff.'

(l) 'The above method can only be used on the parallel portion of the aircraft, further investigation will be necessary before the double curvature could be attempted as this may involve new punches and dies to cater for the varying angles on the frame flanges.'

(m) 'The cylindrical specimen was then lowered into the prepared safety pit and the thermocouples run out to the cold junction which comprised a thermos flask filled with melting ice; copper extension leads ran from the cold junction to the Browns Recorder as shown in the wiring diagram.'

The following two extracts, drawn from one report, are given with no comment at all, other than to say that no specialism, however detailed, can justify such cumbersome sentences.

The E.A.S. corresponding to 270 kts. I.A.S. are below those corresponding to the V_{NO} given in D.H./C.A.A./1 for all altitudes up to 31,300 ft. T.P.A., the A.R.B. having already accepted (minutes of meeting of 4 February) the V_{NO} curve given in the same report (this earlier V_D curve is identical to the current one given in Fig. 1 except that now $M_D = 0.74$ instead of 0.73 true).

Because of the reduced V_C below 6,000 ft. T.P.A., $V_{NO} = 270$ kts. I.A.S. exceeds V_C between 4,500 ft. and 6,000 ft. T.P.A. it has been accepted however that this does not warrant the quotation of a lower V_{NO} for this altitude band if a warning is given in the aircraft's Flight Manual regarding flight at V_{NO} below 6,000 ft. when there is a risk of collision with birds, since the latter is the only reason for restricting V_C to below about 290 kts. I.A.S. in this altitude band.

The following amended versions of the extracts (a) to (m) could no doubt be bettered; they are offered as improvements on the originals. I have tried to keep as near as possible to the style of the original versions to show that improvements can be made even accepting the writing as it is.

(a) 'When single details are ordered, all references to any scheme, other than the one being ordered, must (or should) be blanked out or scored through on sheet 3 of layout. This will make sure that only the information and quantity required for the particular scheme is ordered for the layout.'

This is an instruction. 'Will be' is not emphatic enough. Even better to say 'you should blank out'. The original also contains our old friend the unrelated participle, 'when ordering. . .'.

(b) 'In all cases of chobert rivets not securing stiffener attachment angles, the attachment angle has been bent away from the adjacent structure. This has made it impossible for the chobert rivets to engage in the holes in the attachment angles.'

Why not even: 'In all the examples (we have) tested where the chobert rivets have failed to secure the stiffener attachment angles. . . .'

(c) 'The tail loads produced by any oscillation in the elevator channel, between the limit switches up to five seconds from the start of the oscillation, are small compared with the design tail load.'

The failure in the original was to find at first glance the subject of the sentence, 'the tail loads', and its main verb, 'are small'. The commas help to highlight them.

(d) 'The case has also been considered of the pilot winding on the trim, until the fine limit switch operates flap up or to full extent, and then disengaging the autopilot with flaps down. (The coarse limit switch range is greater than the trim range.)'

Two actions are involved here, surely. The pilot winds, then disengages. The containment of the qualification by commas shows this up. To the layman this sentence, like the others, may still remain obscure, but at least the ideas are being separated into portions which can be taken in.

(e) 'On special jobs where extremely heavy cutting pressures are required – such as the breaching of high-tensile steel – the normal cutting oils are not good enough. Even though it is extremely expensive, this oil is used because it gives the finish required.'

In the original version the main sentence is 'this oil is used'. You have to wait a long time before you get to it. You will see that it is necessary not only to make two sentences but to re-structure the passage.

(f) 'There is, however, some reduction of the warning margin, which is thought to be caused by a loss of lift. The loss of lift is caused by flap blockage if the pilot prolongs the float on landing when the aircraft is very close to the ground. Although, therefore, there is some reduction of the margin, it is still considered to be adequate.'

The original tries to say far too many things. The main point is that, though there is some reduction of the warning margin, it is still adequate, but the reader is distracted by too many qualifying statements, and by images which appear incongruous. There is an absence of subject. What does 'landing' refer to. Float? Attempt? Surely only a pilot can land a plane. The writer must stick to the passive voice with the inevitable impersonal 'if an attempt is made'. There seems to be a contradiction between 'on landing' and 'very close to the ground'. Which is it? It cannot be both. It can, of course, if 'landing' means 'coming in to land' as opposed to 'taking off'. Why not say so?

(g) 'The main cause of the failure is considered to be the exertion of unequal loads around the periphery of the glass whilst the masking frame is being assembled. The unequal loads are caused by. . . .'

What does 'due to' refer to? To have 'cause' and 'reason' in the same sentence is tautologous. Either there is one cause of failure or there are a number of reasons for failure but not both. The reasons must surely be a further explanation why there should be unequal loads around the periphery. One usually, incidentally, expects a colon after the statement 'one of the following'.

(h) 'From what has been seen during this investigation, it is most strongly recommended that the practice of scraping the casting be immediately prohibited (stopped). Other means should be sought to allow for the known discrepancies in casting, such as those suggested in the following paragraph.'

The comma in the first version between 'casting' and 'be immediately' is pointless. The sentence benefits from being made into two. In any case, in the original the second half of the sentence is too closely linked to the first half to justify a semi-colon. 'To allow' is neater than 'making allowance for'. Such compound phrases as 'known casting discrepancies' should be avoided where possible. One of the features of the English language is the fact that it is not an inflected language and that prepositions are used instead of endings; prepositions are fast disappearing from technical writing in favour of long compound noun phrases. The phrase 'in the following paragraph' carries you along with an easier flow than the abrupt decimal 4.3, which at first reading might almost be taken for a discrepancy measurement.

(i) 'At the start, speed and time of contact were fixed, while pilot material, lubricant, and load were varied; once pilot material and lubricant had been determined, load and speed were varied and, finally, time of contact.'

The sentence here is a very good example of one which lends itself to the use of the semi-colon, where there is a balance, a contrast, a reverse of action, like two sides of an equation. The semi-colon gives the statement more shape than a full stop would have done. The rather obvious use of commas further clarifies the meaning and builds up the architecture.

(j) 'A smooth, surface finish which had been produced by either a rubber-abrasive wheel or 400-grade emery paper – however carefully the article was cleaned afterwards – tended to produce scoring.'

The phrase 'however carefully cleaned' gains by separation with dashes. It is not really the 'finish' which is cleaned but the component, the article, the job on hand. This should be inserted for clarity.

(k) 'The reports marked with ticks will be issued from the new Library. Any reports not in the new Library will be immediately transferred from the old Library. All of these, together with a copy of the list, will be taken back to the member of the staff.'

Instructions in procedure, as in instructions on a fire extinguisher, always benefit by brevity. Three sentences are here better than one. 'Any reports' rather than 'any' and 'the new Library' rather than 'it' all make easier reading. I am still not sure what the last sentence means. It is not clear whether the reports still in the old Library are to be transferred to the new Library or taken back to a member of staff. The Library staff? I give up.

(l) 'The above method can only be used on the parallel portion of the aircraft. Further investigation will be necessary before the double curvature can be attempted, as this may involve new punches and dies to cater for the varying angles on the frame flanges.'

Better even than this version would be the reversal of the last sentence, placing the subordinate clause 'as this may involve ...' before the main clause. 'Further investigation will be necessary. ...' 'This' refers to 'above method' and should be placed nearer to it. The tenses should agree, so I have changed 'could be' to 'can be'.

(m) 'The cylindrical specimen was then lowered into the prepared safety pit and the thermocouples were run out to the cold junction. This consisted of a thermos flask filled with melting ice. Copper extension leads ran from the cold junction to the Browns Recorder, as shown in the wiring diagrams.'

The use of the semi-colon by the writer in the original version is flat and therefore pointless.

SUBORDINATION You will have observed that certain groups of words go together, that writing has a certain architecture and symmetry. Sentences can only have one main verb in them unless they are compound ones, but a succession of main verbs would be dull in the extreme. The expressions would be trite and the total effect plain and boring. A child of course can only make simple statements, but gradually as it grows it learns to see certain relationships of time and reason. So that 'I have a puppy. He is a nice puppy. I take him for a walk each day. I give him a bone for lunch,' gives way to 'After I have given my puppy a bone for his lunch, I take him to the park for a romp, where we both have a nice time.'

In order to explain something of some complexity it is necessary to introduce some form of qualification. Most good writers tend not to have more than two or three subordinate clauses in their sentences. The great secret, of course, is variety.

A short sentence at the end of a paragraph containing some longish sentences can be most telling. It requires very much more literary skill to construct an effective long sentence than a short one.

PARAGRAPHS It would be almost true to say that the axiom 'Keep your sentences short' might equally be adapted to 'Keep your paragraphs short', though not for quite the same reasons. Writing shorter rather than longer sentences is in the interests of clarity. Paragraphs nowadays tend to be short for reasons of lay-out and general appearance.

The schools of journalism caution their clients: 'Avoid solid blocks of print; they weary the reader's eye.' I was taught paragraphs at school (and essay-writing) with Macaulay as a model. I had to admire and imitate the broad, sweeping topic sentence, which opened the paragraph and which was followed by something which stretched for the best part of a page. In my efforts to imitate Macaulay I not only wearied the reader's eye but probably persuaded him I was a cretin as well, for I included all sorts of things not particularly relevant to that fine opening sentence.

The pendulum has swung too far the other way, and in newspapers, particularly the popular ones, the paragraph hardly exists as a recognizable unit, for virtually every sentence is a paragraph. All this is in the interests of typography and readability. 'Give the reader plenty of white space,' is the cry. The impact of newspapers has certainly had its influence on general attitudes towards the appropriate length for a paragraph.

Sir Ernest Gowers, who devotes but thirteen lines to the paragraph, sums the matter up best when he says:

The subject does not admit of precise guidance. The chief thing to remember is that, although paragraphing loses all point if the paragraphs are excessively long, the paragraph is essentially a unit of thought, not of length. Every paragraph must be homogeneous in subject matter, and sequential in treatment of it. If a single sequence of treatment of a single subject goes on so long as to make it an unreasonably long paragraph, it may be divided into more than one. But you must not do the opposite, and combine into a single paragraph passages that have not this unity, even though each by itself may be below the average length of a paragraph.

The passive voice and the impersonal approach

The passive voice certainly has a legitimate use, but it is obviously less direct than the active voice and by its nature requires more words to say the same thing, even something as simple as, 'Gravitation was discovered by Newton' instead of 'Newton discovered gravitation'. You may argue, correctly, that a different emphasis is given, and that where the agent or actor is less important than the action the passive voice is preferable.

PSEUDO-OBJECTIVITY A report which is written exclusively in the passive voice, however, does read heavily. One of the main reasons for the use of the passive voice is the correct belief that objectivity is the hallmark of all good reports – with the emphasis on the experiment rather than on the observer – and the incorrect deduction that only the impersonal approach confers this.

This belief is really pseudo-objectivity. If one person has done the research and that person has written it up in a report, the findings are either valid or invalid regardless of whether the writer hides behind such expressions as, 'it was observed that', 'it was further noted', or admits responsibility by saying, 'I observed that' or 'I further noted'.* These, incidentally, are

*There are many variations on the 'it ... that' construction, most of which have a deleterious effect on style:

> It is assumed that ...
> It is apparent that ...
> It has been shown that ...
> It might be thought that ...
> It is to be supposed that ...
> It is considered that ...
> It is proposed that ...

are but a few.

not the only alternatives. A report sprinkled liberally with 'I's' can be irritating and distracting and draw too much attention to the observer. The introductory paragraph of the following report about liaison with another firm has an overdose of 'I's'.

For my first few days I spent my time learning the paperwork system that is used here, and it has been well worth the time. Although I found that I could get things moving by short cuts, I still found that paperwork was an integral part of the job, and had to be correct. To further my education along these lines I took to stores a complete set of paperwork and personally kitted up the Homing Head Amplifier. This again I found was to stand me in good stead with problems that arose later as I now fully appreciated all the facts. I have established good relations with the following personnel and must say that all are making every effort to assist me.

The whole style of writing in that extract seems to be wrong apart from the over-use of the first person pronoun. On the other hand in the next two extracts the manifest objectivity grates almost as much.

(a) 'During Fuel Jettisoning'
'It is intended to state that during fuel jettisoning the speed should not be greater than 200 kts: since V_C in this condition is 205 kts E.A.S. and the p.e. correction to A.S.I.R. is always negative it is considered that this speed should be satisfactory.'

(b) 'In the "Proposed Method" it was first decided to state exactly what path each component should follow through the Cleaning Shop.'

In the first extract both phrases 'it is intended to state that' and 'it is considered that' are unnecessary. If the writer wished to emphasize the fact that the speed should not be greater than 200 kts whilst jettisoning fuel, it would have been better to have said, 'It is very important that during fuel jettisoning the speed should not be greater than 200 kts.'

The second extract is taken from a Work Study report submitted by two engineers. It seems far more appropriate here for the authors to have said:

'In the "Proposed Method" we first outlined what path each component should follow through the Cleaning Shop.'

Another coy method of achieving so-called objectivity, but which deceives no one, is to use expressions such as 'the writer', 'in the author's opinion'. The occasions for using such expressions are not frequent in most reports. Why not, then, say 'In my opinion, we should not purchase these machine tools' or 'I believe the inspection to have been inadequate.' This is a much more direct statement than the impersonal and the use of the passive, and brings the writer and reader into closer contact. The writing becomes more vivid. One advantage of the personal pronoun is that it discourages vagueness and emphasizes a man's responsibility for what he writes.

THE USE OF THE FIRST PERSON The use of the personal pronoun usually occurs when the writer wishes to be emphatic or pass an opinion. It would be both bad taste and monotonous to repeat 'I' throughout a report. A fair guide to work on is that if you are the sole investigator and writer of a report, the use of 'I' is appropriate; if you are one of a team of two or more, 'we' would be correct; if you are reporting the action of some of your subordinates, you can also use 'we' as in the royal plural, but perhaps better in this case to use 'it', e.g., 'whilst carrying out the inspection, it was found that ...'.

You will observe straight away that though it is typical, this sentence is grammatically wrong. In order to avoid the unrelated 'carrying' it is necessary to re-write the passage, 'whilst the inspection was carried out, it was found that ...'.

This re-writing illustrates some of the difficulties to be found with the passive voice, i.e. greater probability of mistakes in grammar, greater likelihood of long prepositional phrases, more circumlocution and more vagueness and more misunderstanding.

You must bear in mind who is your reader. Whereas it might be in order for you to record within the firm your own opinion, if you are dealing with an outside firm, this same opinion may have to be identified with that of the organization for which you work. In that case, if this is the practice of your firm, you should write 'we found that ...', or in the case of a committee, 'the committee felt ...'.

It is also obvious that, by and large, personal opinion will tend to find its expression in the conclusions and recommendations of a report rather than in the body where the emphasis is on the work carried out. Even the conclusions and recommendations need not carry the personal pronoun. Opinions can be expressed directly without the wordy 'it is believed that' or the briefer 'I believe' – merely: 'The customer should not be re-imbursed; the charges he made are groundless.' The reader will know that it is the writer's opinion, particularly if the language is as plain as that.

The best guide as to whether to use a personal reference or not is to be sure that the reader is clear in each case who the agent is. If he might be unsure, it is better to name the agent. In the midst of a research report, say, written by one person and dealing with some analysis that he has carried out previously, if the writer mentions 'a similar analysis was carried out last year', the reader is not to know whether the present writer performed this analysis too, or whether it was undertaken by one of his colleagues. The writer should make this clear.

The use of 'I', whether it is allowed or not, certainly produces a friendly, informal, more relaxed atmosphere, as a comparison of the following extracts will show.

1 (a) 'This report is based on a study of six assembly lines. I have taken most of my illustrations from the one I inspected most recently.'

1 (b) 'This report is based on a study of six assembly lines. The illustrations have been taken from the assembly most recently visited.'

2 (a) 'I used a micrometer to check the accuracy of the tool.'

2 (b) 'For the purpose of investigating the accuracy of the tool a micrometer was used.'

ACTIVE VOICE The personal pronoun and the passive can both be avoided by some re-arrangement and the use of the active voice. Thus,

> I studied the sites.
> The sites were studied.

and   An examination of the sites showed. . . .

You should therefore always try when possible to write in the active voice, whether you use the personal pronoun or not, as the statements you make will be clearer. 'Quality has improved' is better than 'An improvement in quality has been made'. The verb, which is the action word, has been smothered here by turning it into a noun.

If you wish to address people in instructions, the active voice is particularly valuable in asserting attention, as the changes which L.T.E. made in their notices will show:

London Transport Executive Notices, before and after revision carried out in 1949.

DOGS

Small dogs may, at the discretion of the conductor and at owner's risk, be carried without charge upon the upper deck of double-deck buses, or in single-deck buses. The decision of the conductor is final.

WARNING

The London Transport Board cannot be held responsible for failure to adhere to the scheduled times of the buses, nor can they guarantee the running of the services to be as stated, though every effort will be made to maintain them. In inclement weather, on Sundays, certain buses are liable to be cancelled without notice.

SMOKERS

Smokers are requested to occupy rear seats.

DOGS

You can take your dog with you if it is a small one and the conductor agrees. It travels free, but at your risk. If the vehicle is a double-decker, you must both go on the upper deck.

WARNING

You cannot hold London Transport responsible if your bus is late or does not run. London Transport does not guarantee that its services will keep to the time-table or will run at all, although, of course, it will do its best to see that they do.

SMOKERS

Smokers are asked to sit at the back.

The amended versions are shorter, have more sentences, no separation of subject from verb, no vague expressions such as 'failure to adhere to scheduled times' (whoever uses this sort of speech?) and simpler, more familiar words, 'if the conductor agrees' rather than 'at the discretion of the conductor'. The phrase 'In inclement weather' is not only cacophonous but about as phoney as you can get.

Employees are more likely to look at a message and respond if you write 'Will you please' rather than 'it is requested that'.

IMPERATIVE VOICE In an instruction manual, however, the imperative voice is justifiable:

> Unscrew cap from bottle carefully.
> Remove grease from neck of bottle.
> Replace cap.

Similarly, when describing the procedure for an experiment in the laboratory, you can adopt the imperative. Notice how easy the following account is to read – much easier than if the writer had used the passive and written:

'The apparatus is assembled as shown in Fig. 1. Then the drechsel bottles are filled with the appropriate liquids and then the furnace is heated to 750°C.'

The original reads:

Procedure

Assemble the apparatus shown in Fig. 1, fill the drechsel bottles and the bubblers with the appropriate liquids, and heat the furnace to 750°C. Introduce the empty boat into the hot zone of the furnace, pass oxygen at 10 ml. per minute and determine the blank titration for the apparatus. If the carbon dioxide measured is equivalent to less than 0.2 ml. N/10 HC1, after one hour's passage of oxygen the blank can be considered satisfactory and no further heating or purging is necessary.

Pipette into the cooled combustion boat an aliquot of sample equivalent to 4 mg. of carbon but in volume not more than 5 ml. Cover the liquid with alumina granules (5-10 B.S. mesh), place in the cool fore-part of the combustion tube and reconnect the combustion train. With oxygen passing at 10 ml. per minute, cautiously heat the boat with the radiant heater and the bunsen burner until dehydration is complete. About half an hour is usually required for the dehydration. Cool slightly, disconnect from the oxygen supply, and quickly push the boat into the hot zone of the furnace. Pass oxygen over the boat for a further half-hour.

Disconnect the absorbers. Wash each sintered disc inside and out with CO_3-free water adding the washings to the bulked contents of both absorbers. Add five drops of phenolphthalein and titrate with N/10 HC1 until the pink colour is just discharged. Add five drops of bromophenol blue, continue the titration to the end-point and then add 10 ml. N/10 HC1 in excess. At this point immerse the sintered discs from each bubbler in the acid to dissolve any adhering barium carbonate and wash them, adding the washings to the main solution. Boil the solution for three minutes, cool and back-titrate with N/10 N NaOH. Carry out a blank omitting only the sample.

Punctuation

Some people imagine punctuation as a set of rules – and indeed there are a number of rules – which they have only to learn, master, and apply. This approach to punctuation must surely lead to sterile writing. You should not punctuate in order to observe some rule, but because the sentence that you are writing demands that you should in order to make your meaning clear.

Punctuation is used to denote separation of thought, whether it be parallel, parenthetic, or subordinate. It is a means to make the reader read quickly and without ambiguity. It should – and does – take place whilst you are writing, but very often for those who think and write quickly, their mind is full of the thoughts for the next few lines or till the end of the paragraph. In the heat of composition (and in the importance of retaining a particular train of thought) punctuation takes second place. This is no bad thing, provided you remember that the act of reading is different from the act of writing. For better or worse we read between one full stop and the next. Commas provide a momentary resting place for the eye, but it is the full stop that allows us to stop (however briefly) and say to ourselves, 'Yes, I have taken that in, now I can go on to consider the next statement.' So a revision of the manuscript or report should be not, 'how nicely I have said this', or even 'how nicely have I said this?', but 'will this make sense to the reader?', 'Is this sentence too long? Does it require more commas to avoid ambiguity and to provide a short pause or should it, in the light of my knowledge of the reader, be completely re-written?'

If you think on these lines – that is, of what will make reading easy for the reader – punctuation will tend to take care of itself. Think for a moment of yourself as a reader rather than a writer, and ask yourself not 'what *can* I read?' (for with sufficient motivation you can be persuaded to plough through the most turgid prose ever written), but 'what *do* I read?' Your answer to this question is likely to be: a piece of writing with plenty of full stops and white space, with short paragraphs and shortish chapters. Although you may not have thought about it, it is likely that the novels you have sped through have been those liberally laced with dialogue. And a feature of dialogue is that the statements tend to be short. It is long passages of description that hold you up.

The rules, in any case, are only rough guides. Sir Ernest Gowers, who has an admirable chapter on punctuation, is forced to admit, 'About the use of the colon there is even less agreement among the authorities than about the use of other stops', 'The use of commas cannot be learned by rule', and he quotes Fowler on the hyphen to the effect 'that usage is so variable as to be named caprice'. He then remarks very truly that, 'The correct use of the comma – if there is such a thing as "correct" use – can only be acquired by common sense, observation and taste,' and adds, 'Present practice is markedly different from that of the past in using commas much less freely.'

Some very good writers, for example, will use the semi-colon sparingly and the colon not at all; other writers will have a marked preference for the dash. True, the longer the sentence the more punctuation marks will be required, but then many writers compose short sentences. I do not believe that the typical engineer sucks his pen and furrows his brow wondering which punctuation mark he ought to use. I take it that by the time he is in a position of some responsibility he has acquainted himself with the basic rules. Occasional reference to some standard text is sufficient to clear up dubious cases. Scrutiny of the selected extracts in the section on sentence length would suggest that the main problem before the technical writer is not the problem of knowing whether to use a colon or a semi-colon, but of breaking up over-long sentences into smaller and more meaningful units.

In the following three extracts from an article in the *Spectator* the sentences are far from short and so a greater strain is imposed on the writer to punctuate in such a way as to hold the narrative together. Decide for yourself whether you think he has succeeded. Very rarely can technical writers get away with sentences of this length.

Dr Wood is an American who made his study for, I take it, one of those American doctorates, and apart from the fact that he does not know the meaning of the words 'disinterested' or 'prior' (I sometimes get the feeling that I am the only man left alive who does), it does not markedly increase one's confidence in his knowledge and understanding of the British political scene to find him saying on his twenty-fourth page that 'No great stigma attaches to a Labour M.P. who associates with communists or writes for their publications.'

Not only does Mr Wilson appear to be gaily conceding that the Labour Party cannot win an election in the absence of widespread economic hardship (which may well be true; if it *is* true it is one of the most depressing truths about British political life to emerge for decades), but he is also saying that the only reasons that anybody had for turning to the Labour Party before mid-1958 were purely negative ones.

Of course, there is always a tendency for a flush of sympathy to break out for the losers of any battle ('Let's help them on their feet again, and build their bloody fleet again'), but I think, despite the vagueness of some of these new adherents about their reasons for adhering (Rhett Butler was equally vague about his reasons for joining the Army of Confederacy when it had become clear beyond all doubt that the South had lost the war), the Labour Party can take some kind of comfort from it.

The writer has not followed any particular rules of punctuation nor is it possible to deduce any. He has tried to separate or mark off the various parenthetical and subordinate statements from the main statements by a variety of punctuation marks.

In each of these three paragraphs there are two main statements – or in the jargon of grammar, two main clauses – parallel or contrasting. Round them are clustered subsidiary statements of varying types of dependency. It is usual in sentences of this length to separate these subsidiary statements with commas, brackets, or even dashes. The writer, in this instance (the three words I have written are parenthetical and therefore are enclosed by commas), has used commas and brackets but not dashes. He has a fertile mind and no sooner does he have the first part of the sentence written than some apt extension occurs to him. He is overfond of the parenthetical remarks; hence the numerous brackets.

In the first paragraph the two main statements are, 'Dr Wood is an American' and 'it does not markedly increase . . . political scene'; in the second paragraph 'Not only does Mr Wilson appear to be gaily conceding' and 'he is also saying'; and in the third paragraph, 'there is always a tendency for a flush of sympathy . . . any battle' and 'I think'.

It does not help greatly in punctuating to know that some of the dependent clauses are noun-object clauses, some adjectival, and so forth. Common sense and logic are the best guides to follow in punctuation. The test of successful writing is whether your reader can take in the meaning without ambiguity and with ease.

As Fowler says, 'the work of punctuation is mainly to show, or hint at, the grammatical relation between words, phrases, clauses, and sentences . . .'. He adds later on, 'one important use of stops is to express the degrees of thought dependence'. The latter observation is probably the surer guide in modern practice. Modern practice is towards the open rather than closed style of punctuation; fewer rather than more stops. Strictly the following sentence should be punctuated thus:

'He could, in fact, have come, if he had wanted to.'

It is permissible and just as clear in a sentence of such brevity to write:

'He could in fact have come if he had wanted to.'

The parenthetical remark 'in fact' unlike those to be found

in the extracts from the *Spectator* is short; it does not hold up the meaning. The dependent adverbial clause 'if he had wanted to' is likewise short and does not need the comma to differentiate it from the main clause. Incidentally, you might like to judge whether in the first sentence of this paragraph I should have encircled the phrase, 'unlike those to be found in the extracts from the *Spectator*' with commas.

Let us leave the last word with Fowler:

It is a sound principle that as few stops should be used as will do the work. ... Everyone should make up his mind not to depend on his stops. They are to be regarded as devices, not for saving him the trouble of putting his words into the order that naturally gives the required meaning, but for saving his reader the moment or two that would sometimes, without them, be necessarily spent on reading the sentence twice over, once to catch the general arrangement, and again for the details. It may almost be said what reads wrongly if the stops are removed is radically bad; stops are not to alter the meaning, but merely to show it up. Those who are learning to write should make a practice of putting down all they want to say without stops first. What then, on reading over, naturally arranges itself contrary to the intention should be not punctuated, but altered; and the stops should be as few as possible, consistent with the recognized rules. At this point those rules should follow; but adequately explained and illustrated they would require a volume; and we can only speak of common abuses and transgressions of them.

By the word 'stops', Fowler does not mean full stops but the lesser marks such as commas. Full stops should be plentifully used, for this will mean that sentences will be short.

In the Bibliography are listed some books you may consult further in the matter of punctuation.

The right word

You will find many writers who can punctuate correctly, who can structure their sentences to give what they feel is the correct degree of emphasis and yet, somehow, do not achieve clarity. The reason for this defeat is to be found in their choice of words. They use too many or they use words which have been robbed of all meaning, or words which are vague in themselves. (The last point I considered in the Introduction.)

How much clearer writing would be if such expressions as those in the left-hand column were replaced by those in the right-hand column.

It is clear that	Clearly ...
It was noted that if	If ...
It is obvious that	Obviously ...
It is observed that molecules which	Molecules which ...
It has a tendency to	It tends ...
For the reason that	Because ...
Take into consideration	Consider ...
If the improvements meant that	If ...

VERBIAGE Let us look at some actual report-writing examples of verbiage, of words which fill up space but carry little meaning. Some of the examples chosen, as so often happens, reveal more defects than mere verbiage. Their punctuation and word order are also faulty. The following at first sight looks respectable (perhaps that is the trouble with it) and typical:

'Should the supply of agreement forms sent to you not be sufficient to meet your requirements, application should be made to this office for additional copies.'

You must have seen such notices on the firm's notice-board. Would it not have been simpler and more effective to have said:

'If you need more agreement forms, please call for them at the office.'

Technical writers are not the only sinners. A Cabinet Minister in the House of Commons referred to a change in policy in his department as being

'Not a cut-back but a deferment of the acceleration programme.'

This statement was no doubt delivered with that air of authority which lends status to an otherwise meaningless sentence. As Humpty Dumpty remarked to Alice, 'When I use a

word it means just what I choose it to mean. . . . The question
is which is to be master – that's all.'

The Deputy Factory-Planning Engineer seems to be taking
his duties rather seriously when he describes them as:

'It is the official function of the Deputy Factory-Planning
Engineer to assist in all possible ways the implementation of
the instructions and requests of the Factory-Planning Engineer; and also to advise him in all matters relevant to the efficient operation of his department.'

This strikes me as an empty form of words or else rather
pompous. I get no clear picture of what is done.

Consider a few more extracts, together with suggested improvements:

(a) 'In order to avoid congestion in the Library when all the
reports are called in later on members of the staff will be asked
personally, in turn, by the Librarian to return all reports that
they can beforehand.'

Alternative

'In order to avoid congestion in the Library when all the
reports are called, members of the staff will be asked to return
beforehand all reports that they can.'

Nothing substantial has been lost and the message is more
immediate.

(b) 'Unfortunately the latest import cuts leave us barely able
to cover flat sheet requirements and all sales effort has been
called off corrugated Fluflex, orders for this product being met
from gradually decreasing stocks. Unless the licensing position
improves, a very serious position will result on exhaustion of
current stocks.'

Alternative

'Unfortunately the latest import cuts leave us barely able to
meet the demand for flat sheet requirements and all sales effort
therefore has been called off corrugated Fluflex. Orders for
this product are being met mainly from gradually decreasing
stocks. Unless we obtain more licences, we will soon be out of
business.'

'Cover' has a suggestion of pulling a tarpaulin over some flat sheet. The main crime, though, is the complete vagueness which results from the double use of the word 'position'. The last sentence could be even more specific. 'Unless we obtain at least 250 fresh import licences ... the following firms will be short of supplies.' The original sentence conveys nothing and inclines the reader to drop the report into his 'let-it-wait tray'. The cliché 'a serious position' robs the message of all urgency.

(c) 'In Sales Offices numerically adequate outside representation appeared to suffer from lack of the support of suitably trained staff.'

That is a very diplomatic way of putting it, if you like, and certainly will not encourage the reader to take any action. The news is broken so gently. I will not attempt a paraphrase here. The writer was sent abroad to Australia to report on a number of issues concerned with technical sales, import licences, office personnel, and staffing. He maintained he was prevented from being more candid in (c) by virtue of the distribution list on his report. The situation was delicate, feelings might be hurt if the truth was out that the office staff were quite definitely incompetent and frequently letting down the sales representatives. The boss was able to have a more truthful statement verbally of what was going on.

This is no easy problem. There is such a thing as tact as well as truth. The writer can so easily get into this style of writing though, even when the claims of tact are not so pressing.

(d) 'From this test, the writer has concluded that if the packing on the fuselage is faired off to mate reasonably with the underside of the casting contour (apparently presenting no difficulty to this concern), then the distortion factor can be considered negligible as far as this investigation is concerned.'

Alternative

'From this test, the writer has concluded that if the packing on the fuselage is faired off to mate reasonably with the underside of the casting contour – and this operation would present no difficulty to Marshalls – then the amount of distortion is negligible.'

'Concern' in this context is not the best way to indicate the firm involved. 'The distortion factor ... is concerned' is not downright bad writing, but it is woolly and cliché-ridden. Is there some obvious difference between 'can be considered' and 'is'? Is not 'the amount of distortion' more accurate than a 'distortion factor'? 'As far as ... is concerned' is very common and can usually be substituted for a simpler preposition such as 'for'. As the whole report is about 'this investigation' it seems rather pointless to add it here.

(e) 'The above paragraphs have dealt with the progress achieved in respect of the inherent technical qualities of the aeroplane, but a comparison of pre-war and post-war air fares corrected for changes in the value of money, will, I think show an even more striking degree of progress than that shown in Fig. 1.'

Alternative

'The above paragraphs have dealt with the progress of the inherent technical qualities of the aeroplane. A comparison, however, of pre-war and post-war air fares – corrected for changes in the value of money – will, I think, show even more striking degree of progress than that shown in Fig. 1.'

What is added by the expression 'achieved in respect of'? The use of the word 'inherent' is also open to question. What are the qualities of an aeroplane that are inherent apart from its ability to fly, and therefore to satisfy certain structural requirements?

(f) 'Applications from four divisions for financial assistance in development of research programmes were given approval by the board of directors.'

Alternative

'The directors granted four divisions money for research programmes.'

(g) 'The Personnel Department has accumulated statistics regarding the number of new installations currently under construction.'

Alternative

'The Personnel Department has gathered figures on the number of plants now being built.'

(h) 'The development of workers into sections and departments has obviously been made with a view to the ability of the respective managers to keep in touch with their people.'

Alternative

'The development of workers into sections has been made so that managers can keep in touch with their people.'

Try your hand at improving the following three extracts:

'Nevertheless, there is a measure of value in the proposal and on balance, it is considered that it should be implemented provided that monthly ordering strictly pro-rata to the yearly figure is rigidly adhered to, once the figures have been accepted by our Technical Sales Department Overseas.'

'However, work is continuing to take account of varying normal acceleration during the climb, with particular reference to the optimization problem.'

'You will recall recently we were discussing the control of the extruding plant in particular with reference to the plastic covering on wire. In this connexion you indicated that you considered the temperature was the critical condition regarding P.V.C. covering.'

ABSTRACT TAG WORDS All the above extracts have been drawn from reports which have passed through my hands, and their sins are very typical of those to be found in much technical writing. You will have been aware of overworked words, abstract tag-words such as 'nature' and 'condition': e.g. 'The track was in a wet condition', 'The distribution of demand is of a widespread nature', when it would be better to write, 'The track was wet', 'The distribution of demand is widespread'.

Some of these words still have a use in the right context but generally they have lost precision because they have been too carelessly used in the past. Here is a list of some of these words.

Appreciable	Factor	Problem
Case	Facilities	Real
Condition	Implement	Relatively
Certain	Overall	Situation
Consideration	Practically	Tendency

These words and similar ones (very, quite, rather) are particularly frequent in *dictated* reports. However convincing they may sound to *you* when you utter them, they usually say nothing to the reader. Delete them from written sentences in which they occur and the sentence is stronger.

PREPOSITIONAL PHRASES The technical writer also prefers the prepositional phrase to the simpler preposition. The occasional use of these phrases is legitimate – though good writers can manage to do without them – but if they appear frequently the writing becomes dull and lifeless. The following phrases can be avoided, omitted, or the shorter word used instead:

According as to whether	In connexion with
A certain amount of	In relation to
A high (or low) degree of	In spite of the fact that
As far as . . . is concerned	In the case of
At the present time	In the event of
For the reason that	In the majority of instances
For the purpose of	Owing to the fact that
From the . . . point of view	Provides a means by which
Having a value of	Referred to as
Having (or with) regard to	The former (the latter)
In a number of cases	With a view to

Typical examples of the use of these are:

(a) Addition to the sample of a ten per cent solution of ferric chloride indicates the presence of salicylates *according as to whether or not* (if) a purple colour is produced.

(b) Though such a bearing has no geometrical clearance it still has *a certain amount of* (some) flexibility.

(c) Discussion of this question is irrelevant *as far as present plans are concerned* (to present plans).

(d) A sealing liquid may be required *for the reason that* liquids pumped may be abrasive, unstable, or toxic. (If the liquids pumped are abrasive, unstable, or toxic, a sealing liquid may be required.

(e) *From the manufacturing point of view* (In manufacture) the proposed modification will present no serious difficulties.

(f) *In the majority of instances* (Most of) the samples showed a deficiency of potash.

(g) At this stage the temperature in the reaction vessel should not exceed *a value of* 25°C.

It is these expressions that stand in the way of clarity. When you read a typical passage it is, at first glance, difficult to tie down exactly what it is that makes the passage heavy going. The impression you receive is a cumulative one. Analysis reveals that the dirt in the works is a compound of poor punctuation, a predilection for the abstract word or the round-about prepositional phrase.

LATINISMS It is curious that in letters to our friends we do not sprinkle the pages with such expressions as 'utilize', 'facilitate', 'necessitate', in other words, the latinized and longer words of the English language. Yet when we write a report some false sense of awe or solemnity descends upon us so that we regard the less familiar word as being more appropriate. Much of this is unreflective, ingrained habit.

All books on report writing reiterate the same slogans: prefer the familiar to the less familiar word, the concrete (and easily picturable) to the abstract. 'Try to find out' is better than 'endeavour to ascertain'; 'factory town' conjures up a more vivid picture than an 'industrial community'. This does not mean that reports should be reduced to words of one or two syllables, but if expressions such as 'endeavour to ascertain' are dominant in the report alongside 'according as to whether' and frequent appearances of 'nature', 'condition', 'character', 'factor', the total effect will not favour easy reading. The remedy is to picture or visualize what you want to say. This will cut out some of the unnecessary and vague words. Then when you revise your writing ask yourself if all the words contribute to the meaning. If they do not, cut them out.

All these faults one could describe as jargon if by jargon you mean padding, unclear speech, roundabout ways of describing what you want to say.

TECHNICAL TERMINOLOGY The word jargon is sometimes used (and critically) to describe specialist language, a language which is only understood by a narrow circle of readers.

A better phrase to describe the specialist's language is technical terminology. Provided that the specialist is only communicating with another specialist, who is in the same field and therefore familiar with the terms, technical terminology is legitimate. It is the specialist's own type of shorthand. Its use saves pages of elaboration which would be necessary if he had to communicate to wider audiences. Difficulty often arises where the audience is mixed, when the writer is faced with boring some or talking over the heads of others.

The following extract is an example of good or adequate writing:

In certain islands in Melanesia, the population is divided into *patrilineal clans*, and it is usual for two *clans* to occupy one village. These villages have a dual *organization*, but the *moieties* are *matrilineal* and are cutting across a predominantly *patrilineal* system.

The Tungas are divided into *patrilineal exogamous clans*. Frequently two clans are associated into an *endogamous* group. The *matrilineal cross-cousin* marriage is *preferential*, although *patrilineal cross-cousin* marriage is also known, but less well approved. The marriages are often arranged when two individuals are still children.

The most complex feature of the Karadjeri of Australia is their complex kinship system, and the regulation of marriage. In common with other Australian peoples, the tribe is divided into four sections which can be called A, B, X, Y.

The sections A and B form one *matrilineal moiety*, and the sections X and Y another. A further complication however is that the groups A, X and B, Y are separately *endogamous*.

The children of the women of section A belong to group B, and vice-versa; similar exchange of children occurs between women of sections X and Y. It will be seen, thus, that sections A and Y together constitute a *patrilineal moiety*, and so do the sections B and

X together. The groupings of sections into *patrilineal moieties* has certain ceremonial functions.

The system is neither *patrilineal* nor *matrilineal*. A child does not belong either to the section of his father or to that of his mother, but it will be seen from the preceding that the section it belongs to is quite clearly determined.*

And if it is not 'quite clearly determined', the reason lies not in slack or verbose writing but in your unfamiliarity with the terms in italics. The sentences are short, the structure of them is nicely balanced. To the anthropologist the following extract might convey as little as the last quoted one does to the typical engineer:

In certain types of systems there is an *invariant* of motion which is equal to the sum of the *kinetic energy* and the *potential energy* of all the particles in the system. This quantity is the *mechanical energy* of a *dynamic system*, and in a *conservative system* it remains constant.

Here is one more example:

In America H. C. Brown was attempting to synthesize useful high temperature resistant perfluoro-alkyl-triazine polymers by homopolymerization of perfluoroalkylene diamidines or by copolymerization of these monomers with perfluoroalkyl monoadimines.

This I would defend as adequate writing; the words used – though unfamiliar – have a precise meaning. The same cannot be said of the next extract, quoted in *The Lancet* and by Sir Ernest Gowers, which pretends to be specialist but in reality is not. It is also appallingly written.

Experiments are described which demonstrate that in normal individuals the lowest concentration in which sucrose can be detected by means of gustation differs from the lowest concentration in which sucrose (in the amount employed) has to be ingested in order to produce a demonstrable increase in olfactory acuity and a noteworthy conversion of sensations interpreted as a desire for food into sensations interpreted as a satiety associated with the ingestion of food.*

*This extract and the other two in this section marked with an asterisk were taken from Z. M. T. Tartowski's paper on 'Jargon and Technical Terminology', given in a paper to the P.T.I. group.

Show (for demonstrate), cane sugar (for sucrose), taste (for gustation), eaten (for ingested), sharpness of smell (for olfactory acuity) — these could all be substituted. That still would not make the passage readable. The only surgery possible is drastic re-casting.

In some passages it is difficult to be sure how much of the writing is legitimate technical terminology and how much rather wordy writing. One suspects that the following could have been simplified:

On the whole these multiple regressions lend further support to the interpretation that the effective leader's generalized and scientific attitudes towards co-workers complexly interact resulting in an optimum leader-keyman distance. This hypothesized distance may be visualized as the sum of the distances engendered by a generalized attitude (ASO) and resulting from a particular interpersonal attitude (expressed by sociometric choice).

I think you will agree that the next two extracts do not represent technical terminology, but only woolly thinking and an inability to make immediate to the reader what is in the back of the writer's mind:

Since coals vary widely in amenability to flotation, comparison of the performance of different cells in different washeries is not very illuminating.

It has been shown that prior to the initiation of the overt military hostilities the primary effect of the development of increased physical knowledge is to make possible technological manoeuvre, i.e. the threat and counter threat of the kaleidoscopic development and redevelopment of a complex of attack and defence weapons system whose primary mission is to affect the military-economic competition in such a way as to provide an advantage to the attacker or the attacked at the inception of hostilities.*

Field-Marshal Wavell never wrote like that — and I would be afraid of going into battle under the banners of the military tactician who did.

The following extract was from a manual which was illustrating some tape-recording equipment in an exhibition at Earls Court intended for the general public:

The rapidly expanding use of tape recorders focuses attention to this outlet and continuous work proceeds on the development of microphones best suited to the purpose. By reasons of the inherently available gain, and recorder characteristics, it is believed that the ribbon transducer cannot be bettered, but some modification of polar response might show improvement within the usual lightly damped enclosure. From the general line of customer inquiry it seems to be thought that the ideal of a wide frontal acceptance angle with suppressed rear lobe is commonplace design whereas by fact some highly expensive commercial examples of claimed cardioidal pick-up show quite small front to rear discrimination. The problems are very real if excessive insertion loss is to be avoided and, bearing in mind the size of average rooms, it is doubtful indeed that linear discrimination is the desired target.

Even if the above had been addressed to a fellow specialist, I think you will agree that it could have been written with greater lucidity. The structure is poor, and the whole passage is riddled with expressions such as 'it is believed that', 'it seems to be thought that', 'it is doubtful that'. Why not, instead of writing 'From the general line of customer inquiry it seems to be that', record briefly 'Customers seem to think'. The passage contains an unrelated participle 'bearing in mind', careless incorrect expressions such as 'By reasons of' instead of 'by reason of' (or better still 'because of'), 'whereas by fact' instead of 'whereas in fact'. There is a strongly marked preference for two or three words where one would do ('desirable' is better than 'the desirable target'). The sentences are long and ungainly. The Fog Index for this passage must be fairly high.

6. Finding the right style

In the previous sections I have analysed the various faults to be found in bad writing; sentences which are too long, predicates a long way from subjects, modifying clauses not close to words they modify, over-use of the passive voice, strings of nouns and adjectives to modify one noun, predilection for long or vague words.

Achieving a personal style

Style, however, is something more fundamental than merely avoiding some of the faults listed above. Nor, conversely, is it just enriching your language with metaphors, analogy, or carefully contrasted or paralleled sentences, desirable though these may be. Style is an outcome of a relationship between you and a reader.

Take the following two examples, one actually written by a firm in this country and distributed to its customers in lieu of Christmas Cards, and the other from Dr Johnson to James Macpherson, whose book Johnson had unfavourably reviewed:

(A)
Dear Sirs,

CHRISTMAS CARDS

XYZ has recently given consideration to the sending of Christmas Cards and, although it is certainly not without its pleasant side, we have decided regretfully to discontinue the sending of Christmas Cards from this year.

I am sure that you will appreciate that the warmth of the feelings of all in the Division who have had, and are having, dealings with you is undiminished, and that as in the past, but henceforth silently, our good wishes will be with you at this time of year.

Yours faithfully,

(B)

I have received your foolish and impudent letter. Any violence offered me I shall do my best to repel; and what I cannot do for myself, the law shall do for me. I hope I shall never be deterred from detecting what I think a cheat, by the menaces of a ruffian.

What would you have me retract? I thought your book an imposture; I think it an imposture still. For this opinion I have given my reasons to the public, which I here dare you to refute. Your rage I defy. Your abilities, since your Homer, are not so formidable; and what I hear of your morals inclines me to pay regard not to what you shall say, but to what you shall prove.

The first extract is badly written. It has, if you like, a poor style. And it is badly written because the relationship is false. The writer was basically uncertain of where he stood, of what attitude he should adopt. He has fallen back on cliché: 'our good wishes will be with you', 'I am sure you will appreciate that the warmth of our feelings is undiminished'. (You can positively feel them glowing, can't you?) He is fearful lest anybody is left out, so he includes those 'who have had, and are having, dealings' with the Company. Every contingency must be taken care of. He has even managed to insinuate a gloss of religion into the letter, 'but henceforth silently', with its hint of 'Silent Night', or the grotesque spectacle of members of the firm solemnly sitting silently in their desks for two minutes in remembrance of those to whom they would have sent cards.

The whole letter is false from start to finish. Christmas cards go out in such numbers. Why not very simply then:

Dear Sirs,

This year we have decided to discontinue our practice of sending Christmas cards. We should, as ever, like to wish you the compliments of the season.

                                        Yours faithfully,

And individual writers can reserve their real warmth for those individuals for whom they genuinely feel it. Occasions will present themselves – at the end of a letter, for example, or even in the amended version I have suggested – for them to express it.

If the Company had sent out my amended letter, far from losing customers, they would have earned the undying gratitude of those many other firms who also had been eyeing anxiously the growth of the expense account on the increasingly costly goodwill gifts.

Compare such a letter with the terse, emphatic statement of Dr Johnson. Not one word is superfluous. Every word says exactly what it is meant to say. It is well written. Johnson knows what he wants to say ; there is no element of uncertainty here. The style is intensely personal.

Its contents, its tone, its clear thinking, its courage, its determination to call a spade a spade ('foolish and impudent letter', 'cheat' and 'ruffian') all find coherent expression. It would require a little re-arrangement, incidentally, to destroy its impact. Try putting the words differently, 'I still think it an imposture', 'I defy your rage', 'I shall do my best to repel any violence offered me'; leave out the rhetorical question 'What would you have me retract?' or the repetition 'I think it an imposture still'. Do this and the letter loses all distinction. But the words could not have found such apt expression had Johnson been unsure of where he stood, of what he wanted to say, and how he wanted to say it.

Confidence in writing, therefore, is all-important. Authority and responsibility breed confidence. You can be bold in expressing your views; 'Clearly the Company should not purchase this machine', rather than 'It is clear that, allowing for all the considerations in favour that have been put forward, it would be unwise for the Company at the present moment to consider the purchase of such a machine'.

USE OF ANALOGY It is not hard to inject personal touches – even occasional dashes of humour – into your writing. The following extracts look as though they were written by an individual, they bear the mark of someone who had to care for what he was saying. The style in such cases is friendly and relaxed. The writer is not too obtrusive.

A

Oil, as you know, is a liquid lubricant. But there are other substances, solids in fact, which also have lubricating properties. Graphite is one of these. Any pipe smoker will know that the way to slacken a tight joint in the stem of a pipe is to rub an ordinary lcad (graphite) pencil over it. Graphite lubricates in a rather different way to oil; it fills up the irregularities in the surface of bearings and enables them to slip more easily over each other, see Figure 4. But graphite is not the answer. It fulfilled a very valuable function during its hey-day, but something more was needed.

B

Whilst this idea is rather difficult to grasp in theory, there is a very simple practical demonstration. Take a pack of ordinary playing cards and hold them between the palms of your hands. Now rub your palms together. The two outer cards of the pack will stick to your hands, but the cards in between will slide over each other quite freely. In other words your hands are the bearing surfaces, the outer cards are the 'plating' of Molyslip molecules, and the cards in between are the gliding layers of molecules.

C

We further observe that the molecular chains are in no sort of order, but resemble rather bits of string with which a kitten has been playing.

D

These machines fall into two categories termed analogue and digital.

An analogue computer makes use of some physical quantity which can be varied according to the magnitude of the number it represents. A familiar example is the slide rule, where numbers are represented by lengths marked on a piece of wood.

As their name implies, digital computers accept numbers in the form of digits and operate on them by a process of counting. The most elementary form of a digital computer is the abacus or counting board, the invention of which dates back to the very early periods of history.

You may have felt the analogy (Example C) conveyed a rather homely image. Would you say the same of the following passage from the same book, a technical training syllabus?

E

Under normal conditions a quantity of hydrogen gas will consist of many millions of these atoms *linked in pairs*, which are very stable and rather resemble a husband-wife combination, going through life together and only separating when the physical or chemical provocation is fairly violent. And, when once separated, each lonely atom will take the earliest opportunity of settling down again to a quiet 'married life' with the first lonely atom it encounters. We find that oxygen behaves in the same way, and always two oxygen atoms like to pair and go through life as if they were one.

The analogy here is amusing rather than illuminating, but as the training syllabus contains many pages of meaty factual statement about such matters as moulding thermoplastics, vinyl polymers, and copolymers, and injection moulding (all interspersed with formulae), a little light relief or human interest does not come amiss. A chance phrase or comparison may be sufficient to touch the mind in a different way. Again from the same book the following sentence gives the impression that it was written by a person equally at home in the garden as in the laboratory.

F

As it is these combinations which *can* occur which give us the world we know, ranging from a lump of iron to a rose petal or a plastic, let us now consider some of them, taking the simplest first.

In the examples quoted the writers have achieved their effect of intimacy by the use of analogy.

SUMMARIZING PHRASES You will also have noticed in some of the quoted passages such phrases as 'in fact' (Extract A), 'In other words' (Extract B). These phrases are not padding, but useful devices for reminding your reader or recapitulating what has gone before. Similar expressions are 'in short', 'in brief'. They prepare the reader for some summary to follow and are signals he will anticipate.

The following extract is a good example of such a phrase, in this case, 'that is', which carries on the flow of the sentence introducing an elaboration. You will notice that this sentence is a very long one, but because it is well structured and well punctuated you can read it without effort.

These chains of resin are very much shorter than the chains of polythene or P.V.C. we have been looking at before, but, when the material is heated, they behave just as do those of the thermoplastics – that is, they become agitated, the forces between them decrease, and the material consequently becomes softer and flows easily under pressure.

METAPHORICAL LANGUAGE Another way of holding attention is by the use of metaphorical language. Shakespeare used metaphorical language, for raising the emotional temperature, for enriching his thought. The engineer's purpose in using it need not be so lofty, but more practical and mundane, for helping the reader to visualize some technical description. Three-dimensional drawings can be more easily identified if some small descriptive phrase, such as 'bell-shaped', 'pyramid-shaped', or 'shaped like an inverted bucket', is added.

Similarly, an object can be more readily pictured if descriptive adjectives are added, such as, blunt, angular, green, short, rough-faced. Although these additional words may not be strictly necessary to meaning or understanding, they aid the memory. Their insertion can be unobtrusive. 'The operating handles are placed to the left of the panel. They have black, ribbed grips.'

A useful way of varying pace in writing is to insert a question even though you supply the answer. 'The enormous losses in manhours have been thoroughly analysed in this report. What steps should be taken to reduce them? In the first place ...'

REPETITION Another convenient method of rubbing home a point to the reader is to give him the same information in two forms, in a negative and a positive form. 'The components did not reach the stores at regular intervals. Dates of delivery ranged from two to six weeks.' It would have been quite in order to have presented this information by merely observing that 'Components reach the stores at intervals ranging from two to seven weeks'. Repetition can make the point above more emphatic but obviously should not be employed on every occasion.

RHYTHM All the points I have mentioned show the good
writer to be alive and responsive to language, sensitive to its
nuances and rhythms.

Take the style of a real individual:

> Ever since your victory at Alamein you have nightly pitched
> your moving tents a day's march nearer home. In days to come,
> when people ask you what you did in the Second World War, it
> will be enough to say: I marched with the Eighth Army.

This extract from Sir Winston Churchill's speech to the
troops at Tripoli is not only an example of using words and
phrases (such as night, pitching tents, march, home) which
carry a picture, but is a good illustration of the importance
which word order and rhythm have to the flow of the sentence.
It could have been written (or spoken):

> You have advanced rapidly ever since your victory at Alamein.
> When people, in days to come, ask you what you did in the Second
> World War, you will be able to tell them you fought in the Eighth
> Army.

I think you will agree that not only does the expression 'advance rapidly' conjure up very little, but the whole sentence is
slack and has lost its impact. If you care to analyse why this is
so you will see that the suspense of the first sentence has gone
because your interest is not being maintained. To put the major
statement first, 'you have advanced rapidly' makes the other
phrase an anti-climax. More than that, Churchill's sentence,
'you have nightly pitched your moving tents a day's march
nearer home' (irrespective of the impact of the words) has a
rhythm about it and rhythm is not something merely to be
found in poetry. So it is with the second sentence. In the para-
phrased version of mine, the phrase 'in days to come' is a fussy,
parenthetical interruption separating the subject of the sentence
from the verb, to which it is nearest related. Although the expression 'it will be enough to say' may strike you as being just
a trifle unusual it is in the circumstances effective, and enables
the much more powerful '*I* marched' to follow rather than the
looser reported speech 'you will be able to tell them. . .'.

You may feel, and rightly, that this sort of language is more appropriate to *Henry V* and a 'These wounds I got on St Crispin's day' type of utterance, than to the impersonal prose of the model report writer. Nor do I suggest that you cultivate this rather rich style. Simple, unadorned mathematical statement is best for reports. But the Churchill extract does reveal that word order, rhythm, and feel for words are not just a matter of chance and these are some of the features which go to make up a readable style and which give the mark of individual composition.

You do not have to write in the highly coloured prose of Sir Winston Churchill to achieve an effective style. Consider first what you are trying to do: arouse the feelings of men weary with fighting and instil in them some pride in their achievements, or elucidate the workings of a jet engine, as Sir Frank Whittle so admirably does in his autobiography, *Jet*.

The major organs of a turbo-jet engine are: a compressor, a combustion chamber assembly, a turbine, and an exhaust pipe ending in a jet nozzle. Large quantities of air drawn in at a front intake pass through these organs in that order. The flow through the engine is continuous. In the combustion chambers the air compressed by the compressor is heated by the steady combustion of fuel. The compressed and heated gases then pass through the turbine thus providing the power to drive the compressor to which the turbine is connected by a shaft. They then pass along the exhaust duct and emerge from the jet nozzle as a high-velocity propelling jet.

That has not only been well written but well thought out. The passage has unity and coherence and an inevitable logic that takes the reader on from one step in the process to the next. It would be a good exercise for you to close the book and write out your own account of the basic workings of a jet engine.

CACOPHONY The following extract is not unclear but its author, unlike Sir Winston Churchill and Sir Frank Whittle, obviously had very little feeling for the words used and was unable to appreciate their cacophony:

Preliminary results on the effect of Z.D.M.C. on the thermal stress relaxation of a peroxide vulcanizate indicate that the selinium compound is an effective protective agent against oxidative degradation.

'Reveal', 'suggest', 'show' are possible alternatives to 'indicate' and would prevent the clash with the rather ugly, recently manufactured noun 'vulcanizate'. Another feature of the writing, as of so much technical writing, is the prevalence of the long compound word or phrase such as 'thermal stress relaxation'. Writing is more effective with apt use of prepositions to show the correct relationship between related nouns. (One sees such phrases as 'buried tank flammable liquid storage facilities' rather than 'buried tanks for storing flammable liquids'.)

'Complete' does not carry quite the same meaning as 'effective' but it may be near enough. One would have to know the context more closely to know whether 'sure' or 'safe' would be appropriate. The impact of the whole passage though, is of someone breathing in your ear 'vulcanizate, indicate, effective, protective, oxidative' – not exactly mellifluous.

The main difference between the technical and the non-technical writer is in the degree of subordination the technical writer should have towards his subject matter. His personal style should not get in the way of what he has to say. Any injection of personality which interferes with the transmission of his message is undesirable.

That being said, you should use all those aids which the novelist employs: variety of sentence length, metaphorical language (where appropriate), correct subordination of ideas within a sentence, repetition of important ideas in different ways, and an easy flow from one sentence to the next. Conversely you should avoid padding, too much dependence on the passive voice, ungainly sentence structure.

By correcting these faults, by finding apt ways of expressing a thought, by being sensitive to the cadence of the language and the balance and architecture of words in a sentence or a paragraph, you will soon develop a satisfactory style. There is 'a wind – or, at least, a breeze – of change' abroad in industry today, encouraging writers to write more freely and naturally, to adopt a more personal and individual style and to abandon the stilted phrases and artificial jargon of 'Commercial English'. Now is the time to go with the wind.

Achieving the diplomatic style

You may argue that the very last thing you want to do is lay the cards on the table – and this for a variety of reasons.

You may not have formulated in your mind exactly what your relationship is with your reader and therefore what it is you are trying to do. Are you trying to convince him of the necessity for some particular action or are you trying to avoid being pinned down? If you are trying to convince your reader, you should use the personal, direct style of writing; if you are trying to avoid being pinned down your language will be rich with such outworn phrases as 'it should be understood that', 'the performance *compares favourably* with', 'the expense accounts could be *significantly reduced*'. No one can be quite sure what you mean. Euphemisms will take the edge from your writing – such expressions as 'reconditioned' for 'second-hand', 'manufacturers' representative' for 'salesman' or 'commercial traveller' – and will make the reader inattentive to the content of the report. And when he does come to think about what you have said he will be met by 'On the one hand this ... on the other hand that. ...' Your uncertainty of where you stand will certainly be reflected in your style.

Here is an example of the diplomatic style:

Nothing has occurred to alter the view that the use of economic sanctions cannot be ruled out if other means of persuasion and pressure are seen to have failed.

This basically means that 'England may have to impose economic sanctions on, say, Katanga.' But to soften that dreadfully objective-sounding piece of news the author unmasks his battery of negatives. '*Nothing has occurred.*' '*To alter*' (a negative as it were by implication.) '*Cannot be ruled out.*' '*Failed.*' '*Alter the view.*' Well, whose view is this? Won't the author be brave and tell us that it is his own previously stated view. No, that would be too committing, so 'the view' must appear to sound as though it was a generally held view. '*The use of economic sanctions.*' By whom? England? Why not say so, if he means England, as he obviously does. '*Cannot be ruled out.*' By whom? *'Means of persuasion and pressure.'* Again, by whom? The writer? Industry? United Nations? The Government?

The passage is 'correct' English but the effect, if not the aim, of such writing is to make sure that nobody can pin anything on to the writer. Its impact is vague rather than clear. This piece of writing is about politics, but the same style, the same sort of expressions such as 'It has to be said that', can be found in technical reports, particularly those which are controversial and call for action.

You may have decided to use the diplomatic style not because you are a coward and ensuring that no one can pin the blame on you but because this is the way to be effective. It is not a matter of 'Must I be *diplomatic* to keep my job?' but 'Must I be *diplomatic* in order to get the desired result?' Your boss may feel that he is the one who makes the decisions, and is resentful and uncooperative if you usurp his position by adopting too forceful a tone in your writing.

In this case your writing will be a mask for your personal way of writing – but a deliberate, conscious mask. In these circumstances the diplomatic style is justified. You are not uncertain at all of your relationship with your reader; you know it only too well. And in your writing you are being sincere to your concept of that relationship. The best style, one could say, is the truthful style – truth to your original conception of what you are trying to do.

Very often, though, the diplomatic style is used out of sheer habit. Familiarity with so many reports which contain it makes you feel that there is something appropriate about this style in the world of business.

THE PHONEY STYLE This attitude arises out of a false sense of dignity – a feeling that some words are more lofty and fitting. This attitude was very prevalent during the eighteenth century. Even so discerning and catholic a writer as Dr Johnson objected to Shakespeare's using what he called, 'Common and vulgar terms'. Macbeth uses a knife to kill King Duncan. Johnson thought this word smacked too much of the kitchen table. Rapier or stiletto would have been much more appropriate and fitting words for a great play such as *Macbeth*. But the impact of the word 'knife' was more immediate to the whole audience – both groundlings and aristocrats – than a more refined word. In Shakespeare's day they were used to seeing Jesuit priests being hung at Tyburn and then cut with the butcher's knife. So it is that people who end their letters with 'Thanking your goodselves' use such expressions, not because they are unsure of their relationship with the reader – though no one on any terms of intimacy would use these words – but because they feel that business letters demand such phoney and mannered expressions. They would never write to their wives thus. The style is basically insincere, cliché-ridden, and meaningless, which is why I have described it as the phoney style.

SUMMING UP Good style, therefore, is a term applied to a wide variety of writing. Darwin, Eddington, Samuel Johnson, and J. S. Mill all wrote very differently, but with a distinctive style. Good style is one which makes some impact on the reader. The author's personality comes through. Poor style usually refers to writing which is involved, where there is little attempt to structure the writing, and usually where the vocabulary range is limited.

There are no hard and fast rules for good writing. Style is a matter of feeling as well as thinking. You can know all the rules of grammar and observe them and still write in a dull fashion. As a technical writer you should aim to be clear and logical, as the mark of technical writing is clarity and effectiveness. You are not, however, likely to be a person whom other people read with pleasure if you have not some sense of rhythm, some feeling for words, some ability to provide balance and contrast in your sentence structure.

Although style is a subjective thing, it is not like a birthmark, something you keep with you all your life. You have the same control over it as you do over the clothes you wear, the hair style you adopt. You can change it by enrichment or by economy, by following the style of someone whose writing you admire. If you get particular pleasure from a certain author it is worth pausing for a moment and analysing how he gets his particular effects, and seeing if you cannot introduce some of his skills into your own writing. There is nothing more useful in improving your writing than reading widely, and particularly good novels.

C. BAKER

7. Illustrating technical writing

Words are one of the tools of communication but they are not
a comprehensive medium for every aspect of every technical
proposition. They convey something of the mental picture conceived by the author – and leave each reader to form his own
interpretation of the subject. For an unfamiliar subject they
may be grossly inadequate, so he illustrates the work and, with
pictures, introduces facts and features which can be grasped at
a glance or studied in detail. Either way they expound the
author's intentions and provide all readers with the same
picture. They form an oasis in the confusion of a wordy
treatise by affording the reader the simplicity of something he
can see. For visual aids aid visual images. They may be illustrations in a document, pictures on a screen, charts on the
wall, statistical diagrams, or scientific data sheets. All convey,
or help to convey, the technical message.

In this chapter attention is drawn to illustrations in general,
and consideration given to their preparation and production
for average technical publications. Who decides when they
shall be used? And how are they evolved? What form should
they take? And who has the responsibility of producing them?
These are questions which arise in every technical office from
time to time. Sometimes they are trivial matters, sometimes of
major importance. The technical 'know how', the means of
preparation, and the impact of publication may be vital to progress and even safety. The challenge of technical writing and
its associated illustrations cannot be side-tracked.

What is an illustration?

In publications the word *illustration* has a wide, embracing
meaning. It is sometimes said that the consequences of some
particular course of action *illustrate* the conditions that pre-

vail. Such results or conclusions are illustrations in an abstract form, yet they are as effective in their way as any well-drawn perspective in an engineering manual. It is wise to consider this aspect because it illustrates the difference between a picture in a technical document and one included in a work of fiction.

The technical illustration is essentially a means for conveying information and, in the more concrete form considered here, is regarded as something that can be reproduced in some form, with or without accompanying text or verbal matter. It is intended to attract the user and enable him to grasp the message more easily. Its creator has some technical appreciation and his outlook differs from that of the artist commissioned to illustrate a novel. After reading the story the popular artist presents his impressions of characters and scenes as he conceives them. They are the figments of his own imagination and even if they raise the reader's interest they contribute little or nothing to the story.

All illustrations have their shortcomings. Pictures appeal to people and whilst their appeal should be helpful the effect can be misleading. It is easier to look at diagrams than read and understand a difficult treatise and, if the quality and completeness of the treatment are judged from a cursory glance at illustrations, which are themselves but supplements to the text, a sadly inaccurate conclusion may be reached. It is patent that diagrams and pictures used in a publication or lecture must be not only a real part of it but completely relevant to the aspect being reviewed.

Forms and illustrations

If a popular article describes an aircraft, a well-chosen photograph of the machine in flight may be the most apt illustration, but if the emphasis is laid on the technicalities of the engine performance the reader gets a better 'picture' in his mind from a series of graphs. For tables and graphs can both make excellent illustrations – but they should not duplicate or seem to contradict one another. The good illustration tells its story clearly and does not repeat information adequately presented in another or better form. Diagrams of different kinds make useful illustrations for various documents, and visual aids – for depicting quantities, plant output, chemical processes, and so on – can be readily adapted to suit the less informed reader.

Electrical diagrams are in a class of their own. Hours lost in deciphering inadequate presentation may easily outweigh time spent in rearranging the original design drawings to make them more readable to subsequent users. Electrical diagrams may be excellent illustrations or poor sketches, depending more on the form of treatment than the quality of the draughtsmanship.

Some engineers' orthographic drawings provide good illustration material, but for reports, works diagrams, descriptive and servicing manuals, three-view line drawings, preferably in true perspective, are usually more informative. The important detail is emphasized and any irrelevant material omitted. Photographs are more suited to giving a general impression of the object than to providing technical information on the details.

The aim of illustrations

All illustrations should make some impact on those who see them; it may be significant, it may be negligible. It may be what the artist or author intended; it can also fall short of his hopes and even create a false impression. Success depends largely on the artist's grasp of the technicalities and implications of the problem, on his knowledge of the user, and on his appreciation of the environment and attitude prevailing when the document is reviewed.

An illustration or group of illustrations for a publication or instructional set piece should aim at a sequence of targets; attraction, introduction, clarification, and summary; there may be others needed for special varieties of work. Before a man will study a proposition he must be attracted to it in some way. A good illustration can provide just that appeal to make a dull subject palatable or a flimsy commercial project show practical potentialities. Attraction leads naturally to a more detailed examination of the subject matter. The user must be introduced to more technical features with which he should become acquainted, and gradually he will develop an appreciation of the new proposition. He is helped by drawing upon his past experience and applying the knowledge gained to the matter in hand – a dry and wordy process, to be alleviated by informative diagrams. Animated sketches of simple chemical reactions, for instance, remind the trained reader of fundamental principles and, without writing down to a lower level of knowledge, accompanying drawings can be devised to show the molecular structure of the compounds which concern the basis of the message.

The informative introduction dispels ideas that the document is an inevitable time-waster and, in it, illustrations are ready to play their part in clarifying the descriptive material or punching home a commercial feature. The use of illustrations to amplify text is discussed later; the aim of illustrations to retain the user's interest and ease his task is considered now – and must be remembered the whole time the technical work is being written.

With the summary or conclusion of the document the findings or recommendations may be presented in diagrammatic or tabular form – the output graph drawn boldly to indicate the anticipated figure, the block diagram to show production processes, or the works plan to outline the disposition of new plant. An illustration is remembered, and one that is striking and carefully designed reminds the reader of a vital message long after the process of deducing it has been forgotten.

Subjects, users, and the authors and illustrators who prepare documents and visual aids differ from industry to industry. Problems vary with every subject and the suggestions put forward here need adapting to special circumstances. But one vital aspect remains. The primary aim of a technical illustration is to inform; appearance and quality in the artwork and the format adopted are all subservient to this primary function.

How illustrations are used

The foregoing suggests something of the scope of technical illustrations. It remains to see which applications are most

widely used and consider their peculiarities and special requirements.

'Paper work' and publications, which are regrettably prolific
in modern industry, absorb a good many illustrations – perhaps
most of those produced – and the more journalistic forms of
writing, such as articles for the technical Press, need many
others. These illustrations may be divided into groups; casual
sketches produced hurriedly for urgent action, professionally
drawn perspectives with elaborate detail, and so on. To be
good, illustrations must contain all the information the reader
can be expected to make use of – and not more than can be
assimilated with the facilities and time at his disposal. Documents are not always like textbooks which can be dropped and
picked up at will. They may contain information to be pre-

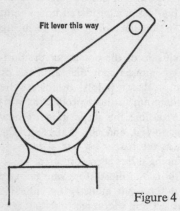

Figure 4

sented convincingly at a meeting or furnish instructions to
perform some servicing operations on an aircraft before its
next flight. Time is the enemy. To meet demands the artwork
must be confined to the essential message and must convey it
clearly. Figure 4 is certainly no picture but it shows, quite
clearly, the way in which a certain lever has to be fitted and is,
therefore, a good illustration. Figure 5 is rather more elaborate
and, provided a descriptive key to the annotations is printed
with it, the illustration is satisfactory as a means of locating
the access panels.

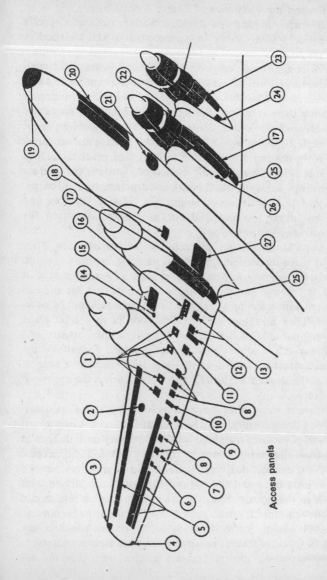

Access panels

Figure 5 A typical physical location diagram

Illustrations for lantern slides, film strips, and similar visual aids demand the simplicity of Figure 4. Seldom can a slide be consulted after it has been screened. So it is useless to include information which cannot be appreciated by the audience in half a minute.

There is a growing tendency to employ illustrations which embrace some form of movement to attract attention and emphasize a fact. Ciné films are universal but the technique of producing them is outside the scope of this chapter. There are, however, many coloured diagrams which give the illusion of movement by adopting suitable lighting on treated surfaces. Both reciprocating and rotating motions are practicable. The technique is ideal for diaramic models of plant and for sectioned drawings showing mechanical construction or fluid flows. Clarity of line, precision of boundaries between surfaces, and good colouring are essential for all illustrations used for demonstration purposes.

A word about models generally is not out of place. They may be three-dimensional, working, or static, replicas of plant, machinery, vehicles, and so on, and made of metal, plastic, or wood, but simple models of cardboard which fold flat and can be issued with a technical or sales brochure should not be overlooked. They introduce a third dimension in its correct plane and provide additional means of simplifying the artistry and conveying information with greater realism. Essentially the models consist of folded cards hinged with adhesive tape to open at the desired angle. The cardboard carries the appropriate pictures.

It has been shown that illustrations for publications are usually part of the text and should be integrated with it. A haphazard selection of pictures sprinkled into the text may do little to help the reader. Illustrations must be carefully schemed and selected to make them useful. Every figure should be mentioned in the text at least once and the noteworthy points established with respect to the details shown. As far as practicable this should be done without laboured reference to pictorial items not immediately obvious or to those on an illustration which is, unavoidably, placed several pages from the descriptive matter.

In some cases, illustrations are used alone; they do not

appear with text and cannot rely on any associated medium to help in conveying the information. Graphs of mathematical functions, temperature charts, and simple mechanical instructions fall within this category. They must truly represent the factors of the proposition or depict the physical make-up of a component, perhaps by exploded views, to obviate the need for unnecessary or prolonged study.

Most technical illustrations are finished drawings prepared in ink, but pencil may be used for quick sketching and artists can produce realistic perspectives which tell the story quickly and naturally.

The users of technical illustrations

Illustrations serve different purposes and are used by different types of people. The form of treatment suitable for a commercial executive is not that appropriate to a technologist. Information for a service engineer is not that written for a prospective customer. To some extent the class of illustration is suggested by the form of writing and the author's direction, but there is a little more to it than that. Take, for instance, specialists' reports presenting the pros and cons of a proposition to executives responsible for making a major decision. Their illustrations should be designed to present the overall picture by diagrammatic or graphic portrayal rather than attempt the presentation of secondary detail which is, more properly, subject matter for an appendix aimed at those who will deal with these specific items.

Information for technicians working away from base, aircraft service engineers or, say, those responsible for the erection and maintenance of plant should contain factual illustrations devoid of irrelevant information with which the users are not directly concerned and be legible and sufficiently 'open' to prevent parts becoming obliterated by contamination with oil or dirt in the field. Not all papers are read in the comfort of the office or classroom. Many are used among machinery, in desert heat and arctic snow – in uncongenial surroundings where error in the message or its interpretation can lead to disaster. The awkwardness of the situation emphasizes the importance of appreciating the user's environment and meeting his needs with illustrations providing information he can use. When the user is remote from the source of information, more care is needed to ensure the completeness of the artwork than is necessary when the recipient is in the adjoining office. Hence the sketches accompanying some laboratory reports used within an organization are scanty. They supply preliminary information for study and the users appraise the work in conference where doubtful aspects can be cleared up.

The same may be said of visual training aids; but teaching really falls into a different category. Students can ask questions,

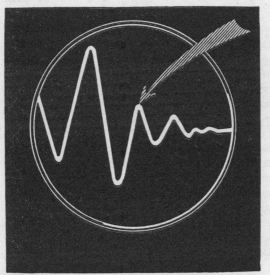

Figure 6 A bold form of negative presentation

but they should not have to ask how to read an illustration. Wall charts and slides, whether diagrammatic or realistic, should be expressive and informative and need the minimum explanation from the lecturer. They should be bold, clearly outlined, and coloured or shaded to differentiate areas, if the drawing is not clear enough in line alone. Size is important. In class, an illustration or projection must be clearly visible from the back of the room, but if wall charts are used for reference in the factory the scale may be much smaller since the picture will be viewed at closer quarters. The poster technique of the commercial artist can be adopted with advantage for some technical diagrams. Attractive designs and lettering are useful memory aids. A white image on a black ground, for instance, is more arresting than normal positive presentation. The curve shown in Figure 6, which might be taken from an oscillascope, is a striking example of the negative technique.

Technical brochures usually carry the best artwork. They are broadly of two types: those intended to advertise a small component in a large competitive market, and those compiled for a specific product or contract, perhaps even for a single prospective customer. Both types require forceful illustrations well printed on good paper. They are intended to influence executives in their evaluation of a tender, and the quality of the work should be governed by the status of the principal recipient. Painstaking artwork is well justified to portray the essential features of the proposition to the man who can make a decision. No one would suggest that a costly, inefficient product could be sold by glamorous picture work, but the publicizing of a sound proposition can be dangerously delayed by technical information confined to freehand sketching on half a sheet of foolscap.

The producers of illustrations

In large organizations the technical publications departments employ specialist artists for each type of illustration: for line work in pencil and ink, for wash drawings and colour reproduced in half-tone, for photograph retouching and airbrush work as well as, perhaps, scraper-board illustrations and screen printing for posters. Within the scope of the present work it is, of course, impracticable to touch on even the principles of basic art as applied to the numerous professional techniques but it is possible to consider the forms of drawing produced both by skilled artists and by draughtsmen or others without art training.

The technical artist has a wider knowledge of engineering than is needed by his commercial colleague specializing on advertising design. He learns to appreciate the production problems of the factory and becomes acquainted with difficulties encountered when working outside. This is necessary because he does not originate information. He gets it from the author or research worker and he must learn 'how to speak their language' in order to understand it. So he should have both technical and artistic training. He learns to put himself in the position of the user and is able to cooperate with the author to interpret the original ideas clearly and accurately.

Professional illustration is usually perspective work in line, monotone, or colour using natural impressions or detail realism in the presentation of plant or products, showing the location and operation of items of equipment or, say, providing illustrations for instructional handbooks. If an accurate or dynamic portrayal of the subject is vital to a proper understanding, good artwork should not be spared. But when an 'amateur' starts drawing, the work and its potentialities are, of course, changed. The inexperienced illustrator may be the technical specialist who started the project. He knows the job thoroughly and he knows what he wants, although it is doubtful whether he knows much of the possibilities of presentation or of the more advanced pictorial techniques. But he can still produce good illustrations, provided he keeps to simple 'flat' or isometric drawings, tables, and graphs which will not betray his lack of artistic skill.

Special attention should be given to matters of reproduction which the specialist illustrator takes in his stride. Line work, for instance, really means line. If it is proposed to print the illustration by letterpress or lithography, tones are only possible by making the masters through a screen which breaks up the picture with dots so, unless this is acceptable, the original must be drawn without pencil or crayon shading, and the lines must be distinct and spaced so that they do not run into each other, especially after reduction. To sharpen up artwork it is usual to draw it $1\frac{1}{2}$ to twice the size that it is to be reproduced.

For reports and memoranda with limited circulations it is, of course, unnecessary to go to print. Reproduction by a duplicating or photocopying process is often acceptable. It may be assumed that these methods need little care, but if the prints are to be reasonable, the originals must be good.

The value and choice of technical artwork

The average technical man appreciates the value of illustrations more than he realizes, and makes use of them more than he thinks. Speak to a draughtsman or technician about ordinary, mundane things and he seems incapable of answering without a pencil! To explain why his car headlight failed or the advantages of taking a secondary road to the coast, he makes a sketch. Sometimes he carries a pad in his pocket to be ready for anyone who speaks to him! Sometimes he relies on the corner of the newspaper. It does not matter as long as he can draw.

Now what promotes this attitude? Is it the lazy way of explaining a thought without having to go through the tiring process of expressing it in precise and simple words? Or is it a way of making it easier for his hearers? It is well known that many readers picking up a book turn first to the pictures, partly because it is customary to do so and partly because illustrations are regarded as a master synopsis from which to judge the contents and treatment of the publication. Of course, neither reason is sound, but since authors and illustrators convince themselves that no users of their work can have sound judgement – a good assumption to inspire better publications! – the viewpoints are worth remembering. The child looks at his picture book and remembers what the hero looks like. He associates him with the story and is reminded of the incidents. The same thing happens when a technical reader studies a treatise. But here it is important for the picture to be accurate in every detail and always to remind the user of the right story. The picture, indeed, must support or be supported by succinct text and the balance between written matter and illustrations equated to the nature of the subject and the cost permitted for reproducing both media. To arrive at the best solution the author must know something about illustrating and the question is, obviously, how much? The application of illustrations and the impact of different forms of treatment on the user are obvious essentials, the need for some knowledge of drawing and photographic techniques is, perhaps, less apparent. The most valuable experience is gained from actually compiling and editing technical copy for publication. If the author is unlikely to be called upon to prepare perspective illustrations himself there is little purpose served by becoming adept at producing them. But he does need some knowledge of the problems connected with drawing to prevent him asking for things which are neither practical nor reasonably cheap to produce. An elementary knowledge of perspective reveals the amount of an object which can be portrayed in a realistic manner without distortion. An appreciation of pencil and ink work shows the degree of detail to be expected from an original drawn on a particular grade of paper. The fine lines of ink show detail lost to pencil and a photograph may look interesting, but obscure the very item required by shadow which only retouching can remove. Then again combinations of line, tone, and colour enhance appearance but the technical improvement, the superiority of the treated drawing to help the user, may be insufficient to warrant the extra cost. Experience of illustration production can decide.

Some illustrations can fulfil several functions. An old figure can seldom be resurrected for some purpose for which it was not intended, but a new one can sometimes be designed to cover several points and thus economies in both drawing and printing will follow. Consider, for instance, a drawing prepared to show the lubrication points on a piece of machinery. It would, no doubt, be in perspective, lightly outlined to show the form and shape of the machine, and have the grease nipples emphasized in solid black or colour. Such a drawing might also be used to show how the whole is assembled, and enumerate the sequence of operations to be followed in doing the work. The two functions are not at variance and there is little likelihood of confusion arising from showing both on one drawing. But there is one aspect which must not be overlooked. Assembly may be done in factory or field and done once only by the erection team. But lubrication is part of repetitive servicing undertaken by different personnel. The descriptive matter for each user is different and will, therefore, appear in different parts of the publication, probably remote from each other. Turning over pages to find a figure is time-wasting and the author has to decide if this can be allowed or whether it would be better to use two illustrations – both prepared from the original layout. The second master would be produced photographically from the first and both finished off by adding their own appropriate notes.

Other factors affecting illustrations

Whilst it is obvious that the primary use of illustrations is to clarify presentation, this is often obscure to the man who first puts pencil to paper. The choice of type depends primarily on the user, his environment, and the mental or physical task he is required to perform, but it is also affected by the cost of production and, to some extent, the number of illustrations used. If there are only two or three, style may be immaterial, but if there are a large number it is desirable for all to follow one or two standard patterns with respect to general appearance, arrangement, style of lettering, and size.

The number of illustrations needed is also important. Clearly, if they are simple charts which give the information without text, the number required depends on the number of topics covered. But how many figures should be used when they clarify text? Should the document become a picture book and look like sheets of comic strip or should it carry the minimum of illustrations? It depends on the subject matter. Specifications and similar papers with a legal slant should not have too many illustrations, especially if they show something which might be interpreted as conflicting with the text. This can occur if the illustration shows anything extraneous to the main subject matter. Specification text should be legally conclusive but an illustration is useful to identify parts of a mechanical assembly by reference to items annotated with a key, as in Figure 5. Statistical documents should not be starved of tables and graphs. It is much easier to absorb figures neatly displayed in tabulated form than decipher material contained in the body of the text matter.

Descriptive work, manuals, and instructional material should, generally speaking, be treated lavishly with informative figures. If reference is made to an actual piece of equipment the reader should be left in no doubt as to what it looks like and where it is. Most of the illustrations are pictorial but there may also be a sprinkling of all the other types.

Presentation of mathematics

An algebraic expression is really a mathematical illustration. If it is of special significance it is usual to display it in a line of its own. This gives something of the status of a figure, but may also be necessary for economical printing. If the material is typeset letterpress the use of horizontal division line takes up a line of type, e.g., $\dfrac{a + b}{2}$. In text matter it is better to use the solidus: $(a + b)/2$, but in displayed presentation this is less important. The clarity gained by isolating the expression from the rest of the text and the retention of the more familiar style compensates for the break in the flow of reading. In the example below, the limits of the definite integrals are set in smaller type which does not detract from the main symbols. The multiplication sign $\times$ is used for numerical and main products and the full point for factorials and so on.

$$T = \int_{r_1}^{r_2} \mu \frac{p}{\varkappa} \times \frac{2 \varkappa \times \varkappa. \, d\varkappa}{2\pi \, (r_2 - r_1)}$$

If an expression is very complicated it may be more economical to set it out on paper and make a block rather than attempt difficult typesetting. It could be hand drawn, which is acceptable for lower-grade work reproduced by duplicating processes, but it is not difficult to use the lettering guides, stencils, or transfers now in common use in the drawing office and studio.

Statistical diagrams seldom need perspective or realistic illustrations but pictorial symbols are useful for presenting information to non-technical readers. For instance, little stick men, crates, and sacks may be used for trade statistics.

Tables

The effectiveness of a table depends on arrangement and the positions of unambiguous headings. The eye follows a vertical column more easily than a horizontal line of figures running across several columns and gaps. These may be too wide for the eye to skip without effort and the presence of any vertical rules

TYPE OF HEAD		BOLT SIZE	GAUGE OF PLATE					
			24 S.W.G.	22 S.W.G.	20 S.W.G.	18 S.W.G.	16 S.W.G.	14 S.W.G.
MUSHROOM	AS 1248C H.T.S.	2BA	560	720	925	1220	1550	1550
	AS 1885C M-S		560	720	925	1070	1070	1070
D. BOLT	AS 1248E H.T.S	¼" B.S.F	–	1040	1330	1770	2360	2900
	AS 1885E							

Figure 7 Part of a badly compiled table

can further hinder the exercise. The possibility of omitting such rules should be considered. It simplifies typesetting of letterpress work but is equally applicable to hand-drawn layouts used for complicated tables and publications which are duplicated. Examine the table partly shown in Figure 7. It is taken from an actual example and has several shortcomings. It gives the strength of certain aircraft bolts fitted through light alloy sheet and includes 'Ultimate Design Load per Bolt' (1) in the main heading, but this is really a heading for the columns listing the loads in the different gauges of plate. Under the column heading 'Type of Head', part numbers and letters denoting the material (H.T.S. and M.S.) are given (2) whilst the type of head (3) is really meaningless since the whole table is devoted to mushroom-headed bolts. The brackets (4) in the bolt size column do not help. It would be better to repeat the '2BA' and '$\frac{1}{4}$" B.S.F. to avoid breaking up the horizontal alignment of the figures. The loads, which are the prime function of the table, come under a main heading 'Gauge of Plate' and 's.w.g.' (or gauge) is again repeated for each column (5). The columns have vertical but no horizontal divisions (6). Now compare this arrangement with that shown in Figure 8. The only vertical line separates the details of the bolts from the load figures. Thin horizontal lines guide the eye from the part number selected to

STRENGTH OF MUSHROOM HEADED BOLTS IN L72 SHEET

| PART No. | MATERIAL | SIZE | ULTIMATE DESIGN LOADS IN PLATES OF DIFFERENT S.W.G.s | | | | | |
			24	22	20	18	16	14
AS 1248 C	H.T.S.	2.BA	560	720	925	1220	1550	1550
AS 1885C	M.S.	2 BA	560	720	925	1070	1070	1070
AS 1248 E	H.T.S.	$\frac{1}{4}$"BSF	–	1140	1130	1770	2360	2900
AS 1885E	M.S.	$\frac{1}{4}$"BSF						

Figure 8 A rearrangement of Figure 7

the figure required and there is no duplication of information. These kinds of improvements are even more pronounced with more complicated data.

Tables usually present the relationship between two variables but, in addition, many of them contain secondary information and reference data. The metric equivalents of British units, for example, may be added in additional columns or, perhaps, the behaviour of one variable under different but allied circumstances. Stressing values for heat-resisting steel, for instance, might be given at high working temperatures and at normal temperatures for comparison.

Secondary information of this kind loses much of its value if printed in a flat, monotone style which makes it difficult to sepa-

rate it from the figures which present the principal information.
The problem can be tackled by using bold face type for the
latter and a light face for secondary information which is not
essential. A similar technique can be adopted for manuscript
tables, but a better form of presentation is to use two colours.
If this is practicable the deepest and brightest tint should be
used for the principal information.

Graphs

Many mathematical functions, trade statistics, and quantity
data can be presented in the universally accepted medium of
graphs. For lucid presentation two factors should be borne in

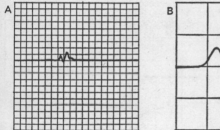
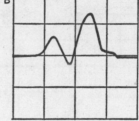

Figure 9 Selection of the best scales for graphs

mind: the scale and the form of graticule. The whole curve
is not always needed; after all some mathematical functions
extend to infinity. So the part relevant to the particular problem
should be drawn to a scale which fills the paper. In Figure 9 (A)
the informative part of the curve is small and overwhelmed by
the graticule, and difficult to read, whereas in Figure 9 (B) the
large scale leaves the reader in no doubt. It will be found that
with a single curve devoid of peaks, such as that shown in
Figure 10, the clearest graph is one in which the scale is chosen
to make the line form an angle of about forty-five degrees with
the axes. Variations in scale can materially alter the appearance
of a curve or family of curves. If the subject is mathematical
and the graph readable the flatness or undulation of the peaks

may be unimportant. But if the subject is statistical the choice of scale can bias the impression one way or the other. Wide variations in the y (vertical) axis can be evened out by decreasing its scale and by increasing the units along the x axis. Emphasis can be given to x or y and this may be justified in some forms of publicity to stress a particular feature.

There is a tendency to plot a graph on any odd piece of squared paper and issue it in the way it is sketched without much thought being given to the user. A printed grid on tracing paper can be reproduced by photocopying methods although it may be faint and difficult to read. By letterpress and litho-

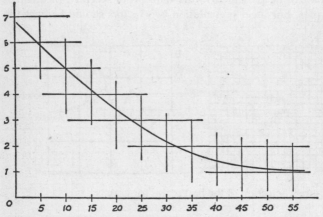

Figure 10 Use of a simplified grid

graphic printing processes the grid will either reproduce as black as the curve itself or it will be lost entirely. The best solution is to print the grid in a light tint such as blue and overprint the curve in black: but this means two runs through the machine and careful register to ensure accuracy. There are less costly ways of achieving the same end. Just as the whole of a curve may not be needed, neither is the whole grid an essential of every graph. If the diagram is drawn on plain paper, the grid can be drawn with thinner lines along the locus of the curve only as in Figure 10. This arrangement not only minimizes drawing work but often improves the clarity of the sheet. Another method by which a light grid can be printed with the curve in a single run is shown in Figure 11. The grey appearance of the graticule is an illusion produced by dotted lines printed in black. Paper of this kind is obtainable in various styles of graticule and is used mainly for making masters for letterpress and lithographic reproduction.

Figure 11 is an illustration intended to serve two purposes, as discussed earlier in the chapter. In addition to showing the grid it displays two familiar curves A and B which, when drawn

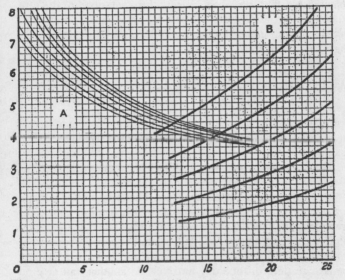

Figure 11 A graph on dotted grid paper

on the same axis in this manner, immediately suggest inappropriate scales and ranges. The curves of family A are too close together to apreciate their significance; those of family B are certainly more readable but, of course, the scale may have been increased at the expense of the range. Some essential information could be cut off the edge of the paper.

There are a few practical hints worth noting in the preparation of graphs. Whilst pencil is satisfactory for a good many

drawings, the presence of a background grid makes it unsuitable
for most graph work. Indian ink is, therefore, recommended.
The axes should be drawn in and the ordinates clearly marked,
remembering that if any annotations fall on the graticule and
the whole is printed in one colour the lettering may be partially
obliterated. When a number of curves are drawn on the same
sheet each may require some form of identification. Annotations printed along the curves must be arranged so that they
do not make the reading of either the note or the curve difficult.
If typewritten or letterpress annotations are used, it must be
remembered that the labels bearing them will obscure the grid
when they are stuck on to the original unless two-colour printing is adopted, in which case the light tint grid is printed over
the whole graph area.

Diagrams

Some graphs, such as those recording rainfall, which consist
of a series of vertical columns, are more aptly described as
diagrams.

This form, which suggests a series of rain gauges, is used for
presenting quantities in which the intermediate readings between two plottings are meaningless. The temperature prevailing over a period may be assumed to have a mean value half
way along the line drawn between readings taken at the beginning and end of the period, but rainfall is in a different category. It is not continuous like temperature. A line between the
rainfall reading on one day and that on another can give no
indication of whether it was wet or fine on any of the interven-
ing days. These diagrams present technical matter in a simple
pictorial form more practical than a graph or table, less realistic
than a drawing or photograph. Other forms of diagrams are
used for electric circuits and adaptations of them are applicable
to pipe-line installations, the layout of an oil refinery, the brakes
of a train, and so on. For easy reading it is essential to adopt a
form either familiar to the reader or one which will recall facts
to introduce the style of presentation naturally. The user should
absorb the technicalities of the problem, unhindered by having

to decipher the form and format of the diagram. Symbols, for instance, should be those commonly used in the industry concerned, and conventions or meanings attributed to thin and bold lines or dotted and chain dotted work should be crisp and self-evident. Once established, standards should be retained without variations.

Just as scale affects the appearance of a graph so the style and treatment affects the impact of a diagram. Pictorial presentation is useful for conveying statistical information, particularly for those who are unimpressed by figures. But the shape of the unit used on the diagram influences the comparison of quantities. In Figure 12 there are pairs of symbols, in each case

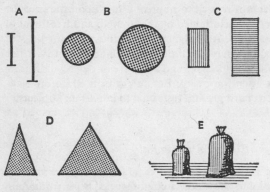

Figure 12 Linear and area symbols to show comparative sizes

the one being double the size of the other. This is apparent with the lines, A ; they are in the same category as the rainfall graph and give an accurate representation of quantity. A line four inches long looks twice as long as one drawn at two inches, but this is not always true of areas. Circles, polygons, and some irregular figures, such as E, may seem less than they really are when compared with a smaller figure of unit area. It may be no disadvantage – in some classes of work it can be helpful. The devices can be attractive and made to appeal to the lay mind, especially if they symbolize the quantity – a sack for merchandise, an hour-glass for time, a workman for labour force, a horse for power. As an alternative to areas devices can be used numerically. One sack, for example, might represent 1,000 tons of corn ; two sacks 2,000 tons ; and so on. This form of presentation acts vividly but one tires of it and, if repeated too frequently, the user gets the impression that the compiler is playing down to him. Not more than two or three such diagrams should be used together.

Diagrams may also be used to identify items otherwise difficult to define ; these may be physical parts such as nuts and bolts, practical factors such as angles or, say, mathematical functions involving the identification of part of a curve such as that shown in Figure 9. Such diagrams should be simple and direct, devoid of unrelated data.

When a diagram is used to impress a fact upon the user the desired impact can be attained by the strength and treatment of the draughtmanship. A straightforward drawing with lines of light but equal thickness has little force behind it, a heavier line treatment is more impressive and a reverse or negative diagram attracts most attention of all. For this reason white on black should be used sparingly and restricted to important subjects.

Electrical presentation

The common forms of electrical diagrams are *Schematics or theoreticals* used to show how a circuit works, *Wiring diagrams* which indicate the connexions between pieces of equipment, *Circuit diagrams* which emphasize the theoretical arrangement rather than practical connexions, and *Location diagrams* which indicate where the equipment is situated. All of them have one essential in common: the need for easy readability. To pursue this end, clarity and simplicity are sometimes achieved by combining one or more functions on one diagram, provided the circuit is sufficiently simple to allow it.

The schematic, showing only essential lines and utilizing known symbols, is the easiest diagram to follow. Figure 13 is an example of a simple circuit drawn in this straightforward manner ; it shows the supplies to an aircraft landing lamp which retracts electrically into the wing.

It is usual to arrange the power supplies, feed terminals, or

busbars at the top or at the left-hand side of the diagram and
start the circuit with fuse or circuit breaker, continue through
the switch gear, relays, or contractors or any control circuit to
the load-consuming item, and then to the return or earth. Sometimes a dotted line is used to enclose parts of the circuit which
form the internal wiring of major components, as is done in
Figure 13. Most circuits are more complicated than this example and the draughtsman must decide whether a diagram
should be split for the sake of clarity. Secondary circuits, for

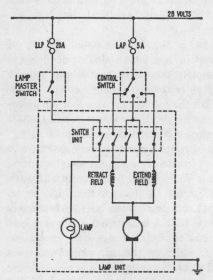

Figure 13 Electrical schematic diagram

instance, should not be allowed to encumber the principal diagram. A main power supply and its control circuit are both
clarified by presenting them on separate sheets. Lines representing cables and connexions should be as short and direct as
possible. On schematic diagrams components may be arranged
to meet this end rather than depict their correct relationship
on the installation. The wiring can also be shortened by avoiding a slavish copy of all physical connexions. For example, a

number of connexions may in fact connect through devious paths to a common earth. But theoretically a series of separate earths may serve the same purpose – and need far fewer lines to show them.

Wiring and circuit diagrams need more detail than schematics since they are used by practical electricians concerned with installing equipment and making it work. Breaks and connexions in the wiring must be shown and the terminals identified. The juxtaposition of components needs some thought but deviations from the correct locations are tolerated, if the actual positions involve excessive 'doubling back' and, therefore, unnecessary lines.

There are various ways for showing the actual location of equipment. The instruments of a radio installation or the components of a set of remotely controlled switch gear can be shown diagrammatically as boxes in seemingly correct elevation. This is satisfactory for the mechanical aspect of the installation but does not, of course, indicate the actual electric circuits since the boxes are only shown connected by cable forms.

Diagrams in the shape of routing charts to some extent serve both purposes. These are frequently arranged as tables using vertical columns; one for each zone of equipment and each break in the continuity of wiring. In Figure 14 the supply comes from circuit breaker, M11, on distribution board AJ, through a plug break pin referenced BFhA to terminal 2 of switch type 6122 situated on the alternator control panel, BF. Connexions from terminals 1 and 2 continue through plug break pins BFbH, and BFbJ, and so on. Sometimes it is necessary to return cables to panels through which some part of the circuit has already passed; this can confuse the wiring and suggests the use of boxes enclosed with dotted lines instead of fixed columns. Boxes can be repeated at different positions on the sheet to avoid the necessity for excessive to and fro wiring.

Routing charts and like diagrams are useful media for communicating special instructions. Vital information can be added to a print of a standard illustration and the amended diagram used as a master for printing the new instructions. Alternatively, the special information can be overprinted in a distinctive colour.

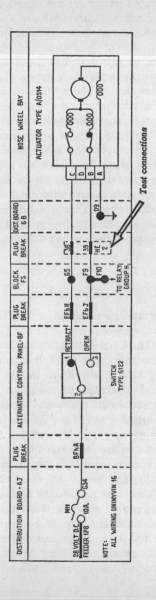

Figure 14 Electrical routing chart

All electrical diagrams call for care in draughtsmanship and layout, especially when reproduced on a smaller scale. A number of lines linked and drawn parallel to one another can become extremely difficult to follow and obscure the salient technical features of the drawing. Lines must be followed carefully with a pencil to see where they go. It will be seen that the verticals are broken to clear those running horizontally. This convention is satisfactory and much better than looping one line over another, a practice which overcrowds congested diagrams. If a complicated circuit is being described it is unwise to confront the reader with a finalized drawing showing all the complexities before the whole has been described and appreciated. It is better to build up the circuit step by step repeating the diagram as necessary with the newly described material added to each view. This is most effective when the additions are made in colour or displayed on transparent overlays which can be turned over by the reader as he proceeds.

Orthographic drawings

Draughtsman's drawings comprise a true plan and elevations of the object, each drawn to scale for design or manufacturing purposes. They carry the essential data needed by the engineers who use them and are fully dimensioned for the works. They may seem confusing to the untrained eye if the subject has several superimposed planes or the viewpoint is unusual, and they are seldom favoured for publications used by those who need a quick grasp of the job and its implications.

When engineering drawings are pressed into service with technical literature they are often taken from works' prints and, if so, contain extraneous detail of construction not applicable to the work in hand; this should be erased. The directions in which views are taken, however, should remain, to avoid doubt as to the form of projection used. It should not be assumed that the reader is skilled in the reading of orthographic drawings nor that he will take the logical and seemingly correct view. He may do nothing more than follow the information actually printed on the drawing and make his own deductions.

Specially drawn orthographic work has none of the inherent disadvantages of workshop prints, and for some subjects is clearer than perspective drawings, particularly plans of buildings, vehicles, and the like. Figure 15, which shows the seating arrangement of a projected airliner, would not give the same informative detail of seating arrangement if it were presented in any other form.

A perspective view would give an impression of the cabin and show the seats but convey different information from the pitch and general disposition of the seating given on the plan.

Isometric drawing

Isometric is a term applied broadly to drawings in which the planes of an object – length, breadth, and thickness – are shown in a single view in contrast to the three separate views – plan, front, and side elevations – used in an orthographic drawing. There are various versions of presentation differing from each other in the angles at which the planes are drawn, but true isometric has an angle of 120 degrees between the planes. No artistic skill is needed in the preparation since dimensions are all measured with an ordinary scale and lines parallel to each other on the object are made parallel on the drawing. Isometric drawings are favoured by engineering draughtsmen and frequently used by technical authors who have no facilities for more professional artwork.

They are often thought of in terms of parallelograms or rectangles, and, indeed, many common objects can be enclosed by an envelope of construction lines forming a cube or rectangular block, but the principle, of course, applies to any shape and if the object contains a variety of planes disposed at different angles a whole series of construction lines may be needed to enclose each plane in its appropriate envelope. A stage is reached when the technician feels that it would be wiser to call in an artist or photographer and get a truly realistic representation of the object. For isometric drawings never really look right. Parallel lines which recede from the observer do not appear parallel, they seem to converge, and if this diminution in

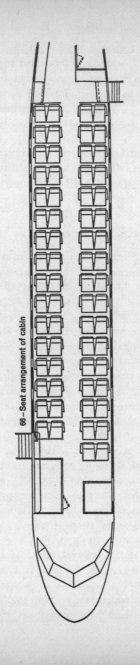

60 – Seat arrangement of cabin

Figure 15 A simple orthographic drawing clearly portrays plan layout

size is not represented the illustration has the appearance of possessing an enlarged background.

Although isometric drawings can be prepared with no more equipment than a scale and a sixty-degrees set square the work is made easier by sketching over a printed grid. There are also a variety of three-view drawing instruments on the market.

The isometric technique is, of course, useful for making illustrations of engineering details and is also suitable for boxes in location diagrams and those representing quantities in lay-out and statistical charts.

Perspectives

A perspective drawing gives the impression of an object as it is seen and so differs from an isometric which combines some degree of realism with geometrical accuracy of the linear dimensions. A good conception of perspective is obtained by viewing a straight railway track from an overline bridge. If the ground is flat enough and the line long enough the rails appear to converge until they meet at a point on the horizon. If the railway had double tracks laid parallel to each other all four rails would meet at the same *vanishing point* but if there was a straight road running by the side of the railway, but not laid exactly parallel to it, the sides of the road would meet at a different vanishing point. Lines parallel to each other meet at a common point, but every set of parallels has its own point.

The sleepers are at right angles to the rails and, perspectively, behave in the same way. Let us suppose that the track is on the left of the observer and that the rails vanish to a point on the horizon which falls to his right; then, if the sleepers were of infinite length, they would all meet at their own horizon vanishing point which would lie on his left.

In making a drawing of the railway track the whole of the construction lines for the rails would form part of the picture from foreground to vanishing point. This is unusual. Only part of the construction lines is normally needed. Most perspective drawings are built up of lines and planes similar to those of the sleepers; they are limited in length and area, and the vanishing point may be well outside the picture area.

Objects in technical illustration are too varied to enumerate, but many of them – machine components, scientific apparatus, vehicles, and so on – can be accommodated within some form of rectangular block in the manner previously discussed for isometric drawings. The observer may be above, or even below, the geometrical ground line and at any distance from the object. All these factors affect the form of perspective construction.

An intelligent selection of viewpoint is a major factor in attaining success. The one chosen must be natural and familiar to the user, easily recognizable, and clearly depict the vital parts of the object.

Perspective drawings can be set out geometrically with the correctly receding lines and diminishing scales by draughtsmen having no artistic gifts, but the process is seldom economic. Accurately laid out work is really the prerogative of specialist illustrators who frequently combine it with some freehand sketching based on a trained eye and the use of various perspective aids. The subject is fully covered in *A Guide to Technical Illustrating* by the author of this chapter.

A form of pseudo-perspective which requires less skill is adopted by some draughtsmen for work involving cylindrical shapes. By showing the edges of end-on circular planes as ellipses instead of lines, the drawing acquires some degree of three-dimensional depth.

Line drawings

The majority of technical illustrations are in line, which is the most lucid form of presentation for scientific and engineering work. Line drawings can be devised to communicate a number of facts and pin-point a particular topic whilst maintaining the environment of the subject as a whole. They may be produced by sketching an actual piece of machinery – this is the easiest way to get good illustrative work and, perhaps, the one which can seldom be adopted. Invariably the demand is for pictures of equipment that has not been made and the only available information is on works and design 'blue prints' or contained in preliminary sketches of the schemes or projects.

Line work is roughed out in pencil and, when sufficient is completed to get a general appreciation of the finished work, an assessment is made of its technical value. This is the time for full collaboration between illustrator and author (if they are not the same person) and a study of the proposed technique. It is useless preparing a drawing which depends on data familiar to the author but unknown to the user or compressing detail into so small a compass that it becomes illegible when reproduced. Artwork must, in short, be *suitable* for the job.

Original pencil drawings can be reproduced by photocopying and some of them by printing, but the best results are obtained from ink originals. These may be produced by tracing the roughs on translucent material or transferring them to white board by rubbing the reverse side of the paper with pencil and tracing through. Ordinary carbon paper is greasy and unsuitable for artwork. Lines should be neither too fine nor drawn too close together. If the style is too delicate the prints may have a broken or filled-in, smudged appearance. Ink work should be firm and the thickness of the lines chosen to compare with the finished result that is desired. Annotations should be clear, both in phraseology and in formation of character, and be so arranged that there is no doubt as to their application. Crowding is inconsistent with clarity and makes reproduction more difficult. The notes given on draughtsmen's drawings are seldom adequate for technical publicity where succinct expression is a feature of the illustrations. Annotations may be hand printed, typed in some form or typeset in letterpress and stuck on to the drawings; they may be stencilled with lettering guides or applied by transfer. The process is unimportant if the results are neat, explicit, and of readable size.

Each illustration should be given a caption and, if it appears in the contents or index, be so worded that the figure is recognizable.

Shading is applied to line drawings to improve appearance, clarify curved surfaces, or emphasize essential features. It may take the form of a firm line, a dot, or an area of solid black. Lines thinner than those used for the outline of the drawing and running parallel to an edge of the surface are among the most effective forms of shading. They may be pitched closer together and merge into heavy black as they run round a curved surface to give a cylindrical appearance. Stippling is quite effective, but the formation of an area of dots with a draughtsman's ruling pen is tedious and the effect is lost if the desired gradation or uniformity of the apparent tone is interrupted by irregular stippling. The best results are obtained by making rows of dots and then crossing them with other rows formed at different angles. The scribbled form of shading is not recommended for depicting hard surfaces. Filled-in solid black, relieved by high-lights in white, is quite effective but, once adopted, the style should be used consistently for all the illustrations in the document and, since it is less easy to handle than other forms of shading, solid black should be chosen with care.

In the studio hand shading is being superseded to some extent by the various varieties of labour-saving techniques now available. The most common one, suitable both for professional and author-produced illustrations, is the use of 'mechanical tints' applied by an adhesive sheet. These tints consist of dot or line patterns and are applied to the surface of the drawing after a backing paper has been peeled off. They are normally supplied as positives, that is, in the form of a black pattern on a white ground but reverse or negative tints are available for drawings made on black paper. The term 'tint' has nothing to do with pigments; the colour of the printed work is, of course, determined by the colour of the ink used in the printing machine. Tints are used normally for letterpress or lithographic reproduction, but they can also be applied to tracings copied by a photoprint process; in this case the adhesive film is applied to the back of the tracing so that it will not adhere to the glass of the photoprinting machine.

Whilst shading improves appearance, it is not always really necessary, and should be omitted from line drawing when the cost cannot be justified or the additional ink could mar the clarity of the prints if subjected to dirt in the workshop.

Line drawings can also be prepared from photographs. One camera process consists of photographing a suitable half-tone original with the use of 'unsharp' masks to give greater definition which, in effect, retains only lines at the boundaries of surfaces and so produces a print resembling a pen-and-ink drawing which can be reproduced by a line block. Another method, used by artists, is to trace the outline of the object in waterproof ink on a photographic print and then remove the image by immersion in a bleaching bath, leaving only the hand-drawn lines.

Half-tone

Natural photographs and half-tone drawings may lack technical detail but possess a degree of realism unobtainable with line. They are best suited to introductory matter designed to arouse readers' interest or record the general appearance of tests, either physical work in the test house or records of dials and instruments.

Originals should have adequate contrast since some of it is lost through the line screen used to make the printing media. They should be free from bluish tinges, which do not reproduce with their visual value, and the detail should be as contrasting as possible. Photographs should be printed on glossy paper and the highlights and shadows emphasized by a trained retouching artist. It assists retouching if the original prints are at least twice the finished size.

Before finalizing half-tone originals, the variety of paper on which the work is to be printed and the coarseness of the appropriate line screen should be discussed with the printer. The wrong combination can ruin the printed work.

Colour

The hues of realism or splashes of contemporary colour please the eye or stir the imagination when skilfully applied to commercial literature. The right treatment has its place in technical publications as well but, for the more serious varieties of scientific document, ornamental colour can create distraction rather than help in appreciating the message. So consideration will be confined to its practical value and the means of treating it.

A pale tinted background or grid is valuable to set off a chart or graph and, if curves are close together, to separate colours, for different colours make the curves easier to read. Choice is important. Apart from being deep and vivid to ensure impact and durability colours should bear some relationship to established convention ; red for heat or danger, green for safety, blue for cooler temperatures, and so on. For export work it should be remembered that foreign customs may differ from those at home and care should be taken to avoid a colour which is, say, a sign of mourning in the country of destination. And again, colour blindness is more common than some authors concede. Colour, and particularly complementary colours, can be an embarrassment to sufferers from this defect.

Colour is effective to differentiate various flows of fluids in pipe lines or currents in circuits and if the convention is maintained throughout a publication an acquired familiarity with the code leads to a familiarity with the subject itself.

Colour can be attractive and often most useful but, invariably, it is expensive ; not necessarily in its original preparation, but in its reproduction. By the usual printing processes each colour requires a separate block or plate to reproduce it and each of these necessitates a separate run of each sheet of paper through the machine. With three colours, including black, the cost can easily be trebled. Colour should, therefore, be used with discretion. It is justified if it will help the reader to understand the work in an appreciably shorter time than would be needed without it. But if the drawing is simple, a figure is limited to a single circuit of cables or pipe lines, or no special warnings or instructions are involved, colour may be nothing more than an extravagant embellishment.

A less costly way of introducing colours into a publication is with tinted paper. An excess of this can give a gaudy impression but the technical value is worth considering if the colours denote chapters of a volume, or technical sub-division such as experimental work, production, maintenance, repairs, or, perhaps more important still, urgent instructions and amendments to loose-leaf manuals issued for the servicing of vehicles and aircraft.

A quick check list

The time to check your report is when you have completed the first draft. Not all writers can enjoy the luxury of a first and final draft, the pressure of time being too urgent. For those of you who wish to stand back from your report at its first stage and measure its likely success, the following check list may be useful. The longer the time you leave between your writing and revision the better, as you can then approach it more critically. If, in addition, you can get a colleague to take a critical look at your report, this will afford a double check. If your report is a long one a good check on its relevancy is to make a short synopsis of it and see how far your report is a true elaboration.

1. Is the purpose of the report clear? The places to look for this are the title, the introduction, the summary.
2. Is the report suitable for its *intended* readers? It is important to distinguish between primary and secondary readers and be sure what are their needs. How will they use the information? Will they accept its assumptions, explicit or implied? Find the explanations adequate? Find the technical level appropriate? Find the non-technical language appropriate?
3. Is the report effective? Does it achieve its intended purpose? Is the conclusion clear and emphatic?
4. *Comprehensiveness* Is everything necessary included? Does the title mislead? Do you keep all your promises?
5. *Relevance* Is everything unnecessary excluded? Is the degree of relevance maintained?
6. *Development* Is the subject developed in an appropriate 'logical' order? (There may be several appropriate orders.) Is the reader informed of the proposed development?
7. *Balance* Are the component parts given appropriate weight?

8. *Arrangement* Is it easy for the reader to see the structure of the report? Easy for him to refer back? Do your headings agree with the table of contents and plan announced in the introduction? Do the headings help to explain the information listed beneath them?

9. *Introduction* Does it introduce? Does it explain the status and scope of the report? Define the limits? Indicate the proposed development? (Too much historical background before the purpose is made clear only annoys the reader.)

10. *Summary* Does it summarize? Is it concise? Adequate? Informative? Is it independent of the report?

11. *Conclusion* Does it conclude? Is it clearly and cogently expressed?

12. *Appendices* Are they necessary? Appropriate? Mentioned in text?

13. *Diagrams* Does each convey its message clearly? Is the association between text and illustration as clear and close as possible? Do the illustrations contain the necessary minimum of explanatory wording? Are the headings precise and informative? (Common errors are captions and numbers not assigned to figures, figures not mentioned in text, figure references appearing too late to be of any help.)

14. *References* Are they adequate? Clearly made?

15. *Format and Lay-out* Are they attractive and pleasant to the eye?

16. *Symbols* Are they conventional when necessary? Well chosen when otherwise?

17. *Mathematics* Is it necessary? Desirable? Appropriate to intended readers?

18. *Vocabulary* Is it simple or too abstract? Too full of meaningless phrases? Are the sentences of reasonable length?

GORDON H. WRIGHT, A.L.A.

Appendix

Fact finding: A brief survey of method and source

The problem of seeking, clarifying, verifying, and evaluating facts, in a world where knowledge is increasing its boundaries, is a complex matter.

Ideally you should have a personal collection of books, pamphlets, and even microtexts, carefully chosen, within easy reach, and simply arranged. When you cannot obtain all the information you need from your personal, or from the firm's, resources, you should avail yourself of the library and reference facilities offered both locally and nationally.

First of all you should set out clearly what you require and follow your search through step by step, exhausting the material for each step before proceeding to the next. It is no use asking for a book on soil mechanics when the information you require deals with the electro osmosis of soils. Equally it is no good asking for information on the electro osmosis of soils without stating that it is required in the context of a building foundation test on clay soils.

BIBLIOGRAPHICAL SOURCES

(a) Books

Guide to Reference Material by A. J. Walford and L. M. Payne, The Library Association, 1959.

This single volume provides a classified, annotated guide to reference books and bibliographies, with an emphasis on current material and material published in Britain. Reference books are defined as encyclopedias, dictionaries, scientific and technical handbooks, data books, histories, directories, year books, atlases, reviews of progress, and abstracting publications.

British National Bibliography 1950–

This includes a weekly list of practically all books published in the British Isles and a monthly cumulated index which gives authors, titles, and subject entries. Quarterly cumulated volumes and an annual volume are also published. From 1962 excludes foreign books published without a British imprint.

Cumulative Book Index

A world list of books in the English language – slightly biased in favour of American publications. Monthly.
Directory of Directories. Published annually.
The Literature of the Social Sciences, by P. R. Lewis, The Library Association, 1960. A guide to reference books and other sources of information.

(b) Periodicals

Ulrich's Periodicals Directory edited by E. C. Graves, Bowker, New York.
A classic guide to a selected list of current periodicals, foreign and domestic. Published every three years.
World List of Scientific Periodicals 1900–50.
A bibliography and union list of periodicals which gives British locations and details of holdings.
Willing's Press Guide Willing, London.
A comprehensive index and handbook of the Press of the United Kingdom of Great Britain, Northern Ireland, and the Irish Republic, together with the principle British Commonwealth, Dominions, Colonial, and foreign publications. The volume also includes lists of newsreels, news agencies, process engravers, and typesetters. This is published annually.
    Make sure that you get the latest edition of these publications.

DEFINITION OF TERMS

(a) Dictionaries

The first step in searching for and presenting information is the definition of terms and an understanding of their scope and context. For definition you should use either a general dictionary or a technical dictionary such as *Chambers's Technical Dictionary*. In some cases you might use a specialized dictionary; an example would be the *Penguin Dictionary of Civil Engineering* by John S. Scott.

(b) Encyclopedias

For more detailed information than dictionaries are able to provide you should use an encyclopedia. These can be general and extensive, e.g., *The McGraw-Hill Encyclopaedia of Science and Technology;* or highly specialized as in the case of *Newnes' Packaging and Display Encyclopaedia* edited by E. Molloy.

(c) Handbooks

Handbooks give an even more detailed explanation and a clearer indication of scope and context. An example would be the *Tool Engineer's Handbook* issued by the American Society of Tool Engineers. This covers all phases of planning, control, design, tooling, and operations in the mechanical manufacturing industries.

More up-to-date explanation, scope, and development will be found by using annual or periodical reviews of progress. There are now many important series available including:

> *Progress in Automation*, Butterworth
> *Progress in Metal Physics*, Pergamon
> *Progress in Non-destructive Testing*, Heywood
> *Advances in Applied Mechanics*, Academic Press
> *Progress in Astronautics and Rocketry*, Academic Press
> *Progress in Semiconductors*, Heywood
> *Annual Report on the Progress of Chemistry*, Chemical Society.

This gives some indication of the range of subjects covered.

SPECIAL LIBRARIES AND
INFORMATION SERVICES

Apart from the public libraries there are many special libraries and information services whose services may be available
free or on a subscription basis. The National Lending Library
for Science and Technology at Boston Spa, which has taken
over the function of the older Science Library Supplementary
Service, is a comparative newcomer to the scene and, providing a twenty-four hour service in the loan of material, will
prove a boon to any technical writer if the material required
is known by the borrower. Moreover, a reference service for
those who can visit the library at Boston Spa is available and
may well eventually provide a service in the North equivalent
to the excellent reference facilities of the Patent Office Library
in London. The Patent Office Library provides very extensive
resources of periodical and other literature in science and
technology, in addition to the unique collection of patent
literature.

Association of Special Libraries and Informa-tion Bureaux (ASLIB)

On a national scale ASLIB is an organization serving sub-
scribing members. Members enjoy an information bureau
which handles industrial enquiries, a book loan service, a panel
of translators as well as an index to translations published
privately or otherwise. In particular, the two-volume ASLIB
Directory provides an excellent source book to organizations
holding special collections of material. In one volume the
library or parent organization is arranged alphabetically by
name, and in the second volume it is classified by subject.

Better library facilities may be available through local cooperative information services, which are generally known by
an acronym illustrating place and scope of inquiries. A few
major examples are:

CICRIS   A cooperative scheme embracing municipal, col-
           lege, and industrial libraries in West London and
           with the headquarters at Acton Public Library.

HERTIS   (Membership by subscription) based upon the libraries of the County Technical Colleges, with the headquarters at Hatfield College of Technology.

HULTIS   Based upon Hull City Library.

LADSIRIAC (Membership by subscription) based upon the Liverpool City Library.

SINTO   A cooperative industrial scheme based on the Sheffield City Library.

Other organizations may be specifically allied to research, where the investigations may lead to the discovery of new products and techniques, or allied to development, where the investigations are largely associated with manufacture and large-scale production. Source books to these organizations are:

Industrial Research in Britain, Harrap, 1962.
Scientific Research in British Universities, H.M.S.O., an annual publication.
Research for Industry, H.M.S.O.

This latter booklet describes in a general way the work of Government and grant-aided research associations.

If you want more specific information about their research association activities or details of the periodicals they issue, you must study their annual reports.

Her Majesty's Stationery Office provides a comprehensive Catalogue Service. It publishes:

(a) A Daily List, issued every day except Saturdays, Sundays, and public holidays. This list is indispensable to all who must ascertain immediately what is being published by the Government from day to day.

(b) A Monthly List, which lists all Government publications issued during the month except Statutory Instruments, is fully indexed, and includes a loose inset with short descriptions of important publications.

(c) An Annual Catalogue, which comprises a bibliography of all Government publications issued in the year except Statutory Instruments and contains an index which is of special value.

There is an International Organizations Publications Supplement to the Annual Catalogue listing publications of the United

Nations and other organizations, sold in the United Kingdom by H.M.S.O.

There are also Sectional Lists. This is a catalogue of current non-parliamentary publications, with a selection of parliamentary publications, presented in separate lists according to the Government Departments sponsoring the publications listed. The series now comprise sixty-four lists which are brought up to date periodically. Issue is free to applicants.

SPECIFICATIONS

Even when you have defined the subject of the inquiry and located a satisfactory source to enlighten you further about it, you may have to refer to a specification relating to material, product, or practices involved.

In England the most important issuing organization for standards or specifications indicating desired properties, dimensions, limits, and methods of testing is the British Standards Institution which began its work in 1901. It is an independent organization supported by a Government grant and subscriptions from members.

British Standards are issued in the following series:

1. General Series, which includes standards on terminology
   BS.1–5,000
2. Automobile materials and parts   BS.5,000+
3. Aircraft materials and components, A – X   BS.5,000+
4. Standards for schools installations   BS./M.O.E.
5. Codes of Practice, C.P.

The annual *British Standards Yearbook* and monthly *B.S.I. News* provide summaries of those standards available and details of new or modified standards.

PERIODICAL LITERATURE – ABSTRACTS

Finally it may be necessary to locate relevant publications providing information on theory or application. Whilst the book and pamphlet are by no means an outdated source, periodical articles are by far the more important source for current information. As the extent of periodical literature grows, so it becomes necessary for the searcher to refer to specialized publications which will indicate the whereabouts and relevance of the article he needs to consult. These 'catalogues', 'indexes', or 'abstract' publications give bibliographic references to the articles. These abstracts are generally arranged in some form of classified sequence to save searching time. In many cases a précis or annotation of the subject content of the article helps the searcher to decide whether it will be useful to refer to the original.

The most important indexes to periodical articles are:
Engineering Index, New York. This may be received in card form by subscription or, more usually, in the form of an annual volume. It indexes over 1,500 major engineering periodicals and gives simple, short annotations.
Applied Science and Technology Index, H. W. Wilson, Co., New York. This is mainly concerned with production problems and indexes about 200 major journals. It is more up-to-date than *Engineering Index* and is issued in monthly parts before final cumulation into an annual volume.
British Technology Index, The Library Association. This is another newcomer, indexing only periodical publications published in the United Kingdom. It is issued monthly with an annual cumulated volume.

Apart from these there are many other more highly specialized abstracting and indexing periodical publications. The following will give some idea of the range:
Chemical Abstracts, American Chemical Society.
Building Science Abstracts, Building Research Station.
Metallurgical Abstracts, Institute of Metals.
Lead Technical Abstracts, Lead Development Association.
Computer Abstracts, Technical Information Co., Ltd.
Solid State Abstracts, Cambridge Communications Co., U.S.A.
International Abstracts in Operations Research, Operations Research Society of America.
Sociological Abstracts, New York.
Index Aeronautics, Ministry of Aviation.

Bibliography

There are a great many books being published on technical writing, variously entitled. Some of them are narrowly confined to instructions on how to write a technical paper if you are a chemist; others range more widely and attempt to say something about the whole problem of communication within business and industry.

Most of these books come from America. Many of them deal with the most elementary points of grammar – points which are normally picked up by an average fifth-form grammar school child. These American books tend to be both more categorical and systematic than ones published in this country. Some of them are obviously intended as a textbook to be worked through by tutor and students throughout (believe it or not) a thirty-two period session.

In some respects American practice is different from our own in such matters as modes of address, letters of transmittal, so that many of the illustrations inevitably do not make their point with the same force as they would if they had been drawn from British sources. It is not a matter of chauvinism, therefore, which makes me say that many of these books are not always very suitable for English audiences.

The most useful books to consult and ones which in one way or another I have drawn upon in compiling this book are:

BAKER, C., *Technical Publications: their Purpose, Preparation and Production,* Chapman and Hall, London, 2nd impression, 1955.

A Guide to Technical Writing, Pitman, London, 1961.

A Guide to Technical Illustrating, Pitman, London, 1963.

BROOKES, B. C., *Technical Writing in Industry* and *Style Manuals for Research Departments* (two Group Papers, D. G. 18 and 29, read to P.T.I. Group Meeting and privately printed).

CHAUNDY, T. W., and others, *The Printing of Mathematics,* Oxford University Press, London, 1954.

CHISHOLM, CECIL, (ed.), *Communication in Industry,* Batsford Business Publications, London, 1957, (2nd edition).

FOWLER, H. W. *Modern English Usage*, Oxford University Press, London, 1930, (3rd edition).

FOWLER, H. W. and F. G., *The King's English,* Oxford University Press, London, 1930, (3rd edition).

GODFREY, J. W. and PARR, G., *The Technical Writer,* Chapman and Hall, London, 1960, (2nd edition).

GOWERS, SIR ERNEST, *The Complete Plain Words,* H.M.S.O., London, 1954.

GUNNING, ROBERT, *The Technique of Clear Writing,* McGraw-Hill, New York, 1952.

KAPP, R. O., *The Presentation of Technical Information,* Constable, London, 1948.

The First Draft, paper read to P.T.I. Group Meeting, October 1953, (privately printed).

QUIRK, R., *The Use of English,* Longmans, Green, London, 1962.

RATHBONE, ROBERT R. and STONE, JAMES B., *A Writer's Guide for Engineers and Scientists,* Prentice-Hall, Englewood Cliffs, N.J., 1962.

TREBLE, H. A. and VALLINS, G. H., *An A.B.C. of English Usage,* Oxford University Press, London, 1936.

WALDO, W. H., *Better Report Writing,* Reinhold Publishing Corporation, New York; Chapman and Hall, London, 1957.

WILLIAMS, G. E., *Technical Literature,* Allen and Unwin, London, 1948.

Some useful and inexpensive guidance is available from the professional publishing bodies on the preparation of technical papers. These include:

ROYAL SOCIETY, *General Notes on the Preparation of Scientific Papers*, Cambridge University Press, 1950.

THE INSTITUTE OF PHYSICS, *Notes for Authors,* Institute of Physics, 1960.

THE CHEMICAL SOCIETY, *The Presentation of Papers for the Journal of the Chemical Society,* Chemical Society, 1952.

Index

MORE ABOUT PENGUINS
AND PELICANS

For further information about books available from
Penguins please write to Dept EP, Penguin Books Ltd,
Harmondsworth, Middlesex UB7 0DA.

In the U.S.A.: For a complete list of books available from
Penguins in the United States write to Dept CS, Penguin
Books, 625 Madison Avenue, New York, New York 10022.

In Canada: For a complete list of books available from
Penguins in Canada write to Penguin Books Canada Ltd,
2801 John Street, Markham, Ontario L3R 1B4.

In Australia: For a complete list of books available from
Penguins in Australia write to the Marketing Department,
Penguin Books Australia Ltd, P.O. Box 257, Ringwood,
Victoria 3134.

Wordpower

Edward de Bono

Could you make an *educated guess* at the *downside-risk* of a *marketing strategy*? Are you in the right *ball-game*, and faced with a crisis could you find an *ad hoc* solution?

These are just a few of the 265 specialized words – or 'thinking chunks' – that Dr de Bono defines here in terms of their usage to help the reader use them as tools of expression. So the next time an economic adviser talks about cash-flows or the local councillor starts a campaign about ecology you know what to do. Reach for *Wordpower* and add a 'thinking chunk' to your vocabulary.

The Complete Plain Words

Sir Ernest Gowers

Sir Ernest Gowers wrote *The Complete Plain Words* as a result of an invitation from the Treasury, which was concerned about the prevailing standards of official English.

Apart from two chapters on grammar and punctuation, this excellent guide is wholly concerned with 'the choice and arrangement of words in such a way as to get an idea as exactly as possible out of one mind into another'.

It is not only for civil servants that clarity of expression is important. It is equally vital in dictating a business letter, preparing a minute, writing an advertisement, or reporting a crime. Without a trace of pomposity or pudder, *The Complete Plain Words* gently administers the medicine we all require. Reconstructed from the material in two earlier booklets, this reference work can rank with such authoritative books as *The King's English* and *A Dictionary of Modern English Usage*.